Environmental Impact Assessment

Other Meridian titles:

Howard Bridgman, Robin Warner, and John Dodson,
URBAN BIOPHYSICAL ENVIRONMENTS
David Chapman, *NATURAL HAZARDS*
Arthur and Jeanette Conancher, *RURAL LAND DEGRADATION IN AUSTRALIA*
Robert H. Fagan and Michael Webber, *GLOBAL RESTRUCTURING: THE AUSTRALIAN EXPERIENCE*
Dean Forbes, *ASIAN METROPOLIS: URBANISATION AND THE SOUTHEAST ASIAN CITY*
Clive Forster, *AUSTRALIAN CITIES: CONTINUITY AND CHANGE*
Iain Hay, *COMMUNICATING IN GEOGRAPHY AND THE ENVIRONMENTAL SCIENCES*
Jamie Kirkpatrick, *A CONTINENT TRANSFORMED: HUMAN IMPACT ON THE NATURAL VEGETATION OF AUSTRALIA*

Environmental Impact Assessment

Procedures, Practice, and Prospects in Australia

Nick Harvey

Series editors
Deirdre Dragovich
Alaric Maude

Melbourne
OXFORD UNIVERSITY PRESS
Oxford Auckland New York

OXFORD UNIVERSITY PRESS AUSTRALIA

Oxford New York
Athens Auckland Bangkok Bogota
Bombay Buenos Aires Calcutta
Cape Town Dar es Salaam Delhi
Florence Hong Kong Istanbul Karachi
Kuala Lumpur Madras Madrid Melbourne
Mexico City Nairobi Paris Port Moresby
Singapore Taipei Tokyo Toronto Warsaw

and associated companies in
Berlin Ibadan

First published 1998

National Library of Australia
Cataloguing-in-Publication data:

Harvey, Nicholas, 1952–.

Environmental impact assessment: procedures, practice and prospects.

Bibliography.
Includes index.
ISBN 0 19 553809 9.

1. Environmental impact analysis — Law and legislation — Australia — States. I. Titles.

344.94046

Edited by Michele Sabto
Indexed by Russell Brooks
Cover photo by Nick Harvey 1990
Typeset by Superskill Graphics Pte Ltd, Singapore
Printed by McPherson's Printing Group, Australia
Published by Oxford University Press,
253 Normanby Road, South Melbourne, Australia

Foreword

In recent years Australian geographers have produced some excellent textbooks that reflect the maturity of geographical research in this country. However, there is a continuing need for relatively short, low-cost books written for university students, books that fill the gap between chapter-length surveys and full-length texts, and that explore the geographical issues and problems of Australia and its region, or present an Australian perspective on global geographical processes.

Meridian: Australian Geographical Perspectives is a series initiated by the Institute of Australian Geographers to fill this need. The term 'meridian' refers to a line of longitude linking points in a half-circle between the poles. In this series it symbolises the interconnections between places in the global environment and global economy, one of the key themes of contemporary geography.

The books in the series cover a variety of physical, environmental, economic, and social-geography topics, and are written for use in first- and second-year courses where existing texts and reference books lack a significant Australian perspective. To cope with the very varied content of geography courses taught in Australian universities, the books are not designed as comprehensive texts, but rather as modules on specific themes, to be used in a variety of courses. They are intended to be used either in one-semester courses or in one-semester components of full-year courses.

Titles in the series cover a range of topics representing contemporary Australian teaching and research in geography—for example, economic restructuring, vegetation change, land degradation, cities, natural hazards, and urban biophysical environments. Future topics include gender and geography, cultural geography, and coastal management. Although the emphasis in the series is on Australia, forthcoming publications in the series will include occasional titles on South-East Asia, drawing on the considerable expertise developed by Australian geographers in relation to this region.

This book, *Environmental Impact Assessment: Procedures, Practice and Prospects in Australia*, is the ninth title in the series. It systematically examines the environmental impact assessment (EIA) process in Australia, comparing and evaluating procedures in each of the states and territories. It begins by outlining the legislative and administrative basis for EIA in each jurisdiction, and then follows the steps in the EIA process, from the determination of whether an EIA is required for a proposed project to the provisions for monitoring the impact of a completed project. How the process works in practice is then examined and evaluated in seven fascinating case studies drawn from around Australia. Commonwealth EIA legislation and procedures are also explained, and are illustrated with a case study of the contentious EIA process for the Sydney Airport third-runway project.

Throughout the book the author's aim is to explain and evaluate the stages and procedures that are followed in undertaking an EIA and reaching a decision on a project. EIA is a complex process, partly because it is based on legislation and administrative procedures that are specific to each jurisdiction. The author has been able to reduce the complexity of the nine different EIA systems in Australia to a standard set of procedures, and in the process he evaluates how well each jurisdiction deals with each step in the process. The book will therefore be of value to students in geography and environmental management courses, as well as to non-students with an interest in environmental issues.

Nick Harvey is well qualified to write this book. He has over 20 years of experience working on environmental issues, with a particular interest in coastal processes, coastal management, and environmental impact assessment. In the field of EIA he has contributed to environmental impact statements, worked in the Environmental Impact Assessment Branch of the South Australian Department of Environment and Planning, taught and supervised students, and undertaken research. He therefore writes from a wealth of both practical and academic experience. In addition, he brings to the topic a sound knowledge of EIA procedures in other countries, which enables him to provide a comparative perspective on Australian procedures and practices.

Deirdre Dragovich, University of Sydney
Alaric Maude, Flinders University

Contents

List of tables

List of figures

List of legislation

COMMONWEALTH

AUSTRALIAN CAPITAL TERRITORY

NEW SOUTH WALES

TASMANIA

VICTORIA

WESTERN AUSTRALIA

UNITED STATES

EUROPE

INTERNATIONAL AGREEMENTS

Acknowledgments

First, I would like to thank Megan McCarthy for her invaluable work as a research assistant on the background material for this book. I gratefully acknowledge all the various Commonwealth, state, and territory EIA practitioners within various government sectors who have contributed by providing information and updating important aspects of the material on EIA legislation, regulations, and practice, and have also provided insights into particular EIA case studies relating to their jurisdictions.

I thank my former colleagues from the EIA branch of the South Australian government, where my seven years experience as a government EIA practitioner provided a good background for bringing the practice of EIA together with the theory of the EIA process. I also appreciate a number of recent discussions on the subject of EIA that I had with academic colleagues from Australia, New Zealand, and the United Kingdom.

Alaric Maude, one of the Meridian Series editors, deserves special mention for encouraging me to write this book in the first place and for providing useful comments along the way. I also thank my partner, Meredith Hugo, for her support during the lengthier-than-anticipated writing process.

The following people and institutions also kindly agreed to give permission to reproduce photos, diagrams, or textual materials, and their assistance and cooperation is gratefully acknowleged: Tindal Airbase proposal map, copyright permission Commonwealth Department of Defence, courtesy of Executive Officer, 322 Airbase Wing, Tindal: Mt Lyell mining proposal site, courtesy of Copper Mines of Tasmania Pty Ltd; Sydney third runway proposal map and photograph of Sydney third runway site, copyright permission courtesy of Sydney Airport; photograph of Sydney Harbour Tunnel proposal, copyright permission courtesy of Qasco (New South Wales) Pty Ltd; profile of Sydney Harbour Tunnel, copyright permission courtesy of Sydney Harbour Tunnel Company; Magnetic Quay proposal, copyright permission courtesy of McIntyre & Associates.

Nick Harvey
Adelaide, November 1997

Abbreviations

ACT	Australian Capital Territory
AGPS	Australian Government Publishing Service
ALRM	Aboriginal Legal Rights Movement
ANZECC	Australian and New Zealand Environment Conservation Council
CCNT	Conservation Commission of the Northern Territory
CCSA	Conservation Council of South Australia
CEPA	Commonwealth Environmental Protection Agency
CER	consultative environmental review (Western Australia)
CFMEU	Construction, Forestry, Mining and Energy Union
DASET	Department of Arts, Sport, Environment, and Territories
DASETT	Department of Arts, Sport, Environment, Tourism and Territories (Commonwealth)
DEST	Department of Environment, Sport, and Territories (Commonwealth)
DR	development report (South Australia)
EAR	environmental assessment report (Queensland)
EC	European Community
EES	environmental effects statement (Victoria)
EIA	environmental impact assessment
EIS	environmental impact statement
EMP	environmental management plan (Tasmania)
EPA(ACT)	Environment Planning and Assessement section of the ACT Planning Authority

EPA(SA)	Environment Protection Authority (South Australia)
EPA(WA)	Environmental Protection Authority (Western Australia)
ERMP	environmental review and management program (Western Australia)
ESD	ecologically sustainable development
GIS	geographic information system
Green Book	non-statutory guidelines for EIA procedures (Queensland)
IAIA	International Association for Impact Assessment
IAR	impact assessment report (Tasmania)
IAS	impact assessment study (Queensland)
LMAHC	Lower Murray Aboriginal Heritage Committee
NEPA	National Environmental Policy Act 1969 (United States)
NT	Northern Territoy
OECD	Organisation for Economic Cooperation and Development
PA	preliminary assessment
PER	public environmental report (or, in Western Australia, public environmental review)
pers. comm.	personal comunication
PrER	Preliminary Environmental Report (Northern Territory and Victoria)
RAAF	Royal Australian Air Force
REF	review of environmental factors (New South Wales)
SEA	strategic environmental assessment
SECIS	social and economic and community impact statement (Tasmania)
SEE	statement of environmental effects (New South Wales)
SEPP	State Environment Protection Policy (New South Wales and Victoria)
TOC	threshold of concern
UNEP	United Nations Environment Programme

1

Introduction

> **Environmental impact assessment, as a national instrument, shall be undertaken for proposed activities that are likely to have a significant adverse impact on the environment and are subject to a decision of a competent national authority.**
>
> Principle 17 of the *Rio Declaration on Environment and Development*, (UN 1992)

This international recognition of the importance of environmental impact assessment can probably be attributed to two significant factors. First, there has been an increased global environmental awareness that we are no longer living within our ecological means. Second, there is a general acceptance of the need to reduce the impact that development is having on the environment. An important tool in producing environmentally responsible development is the process of environmental impact assessment (EIA), which investigates and assesses the potential environmental impacts of proposed development actions.

In Australia there has been a growing awareness of environmental problems, to the extent that the environment has been one of the dominant political issues during the 1990s. The importance of the environmental debate in Australia stems from major international environmental conferences and reports, such as the *World Conservation Strategy* (IUCN 1980) and the *Brundtland Report* (better known as *Our Common Future*) (World Commission on Environment and Development 1987). In the lead-up to Australia's participation in the United Nations Conference on Environment and Development conference in Rio de Janeiro, three important documents were prepared: *Intergovernmental Agreement on the Environment*, *A National Strategy on Ecologically Sustainable Development*, and *A National Greenhouse Strategy*. Each of these documents (all of them were produced in 1992) recognises the importance of the EIA process, but the *Intergovernmental Agreement on the Environment* goes one step further and sets out a separate schedule for EIA in Australia.

So what is EIA? As mentioned above, it is a 'process' to investigate and assess potential environmental impacts of proposed development actions. However, this is a very broad definition that does not say anything about the type of environmental impacts or development actions involved in EIA, or even what happens to the investigation and assessment once they have been completed. In fact, there are numerous definitions of EIA. Some refer only to the biophysical environment, others include social impacts, and others detail the scope or purpose of the assessment. There are also different definitions of the term 'environment', and there are even specific legalistic definitions of what is meant by 'development'. For the purposes of this text, the following definition of EIA is used:

> Environmental impact assessment is a process of identifying and predicting the potential environmental impacts (including bio-geophysical, socio-economic and cultural) of proposed actions, policies, programmes and projects, and communicating this information to decision makers before they make their decisions on the proposed actions.
>
> Harvey 1995, p. 1

Although this definition includes policies and programs (compare Munn 1979), these have often been ignored in EIA studies so that today there is increasing focus on what is called strategic environmental assessment (Therivel et al. 1992; Therivel and Partidaro 1996). Strategic environmental assessment examines policies, programs, and plans rather than focusing on the project at issue. There are also other areas of impact assessment studied in their own right, such as social impact assessment, risk assessment, and environmental health assessment. A number of recent texts produced by EIA research centres in the

United Kingdom provide an excellent introduction to EIA (Glasson, Therivel, and Chadwick 1994), outline EIA methods (Morris and Therivel 1995), and provide a comparative review of EIA in different countries (Wood 1995). There are also less detailed texts on EIA recently produced by Australian-based authors such as Gilpin (1995) and Thomas (1996). It should be noted that in the United Kingdom legislation on EIA is fairly recent: it was implemented in 1988 following a European Community (EC) directive on EIA in 1985. For this reason United Kingdom researchers (such as Glasson and Wood) have shown a research interest in Australian EIA, which, in contrast with the United Kingdom, has a history going back to 1974, although the origins of EIA itself go back even further.

THE EVOLUTION OF EIA

EIA procedures were first formally introduced in 1969 as part of the US *National Environmental Policy Act 1969*, which was signed into law on 1 January 1970. Subsequently, EIA has been adopted in various forms around the world, with the notable exceptions being a number of African and Middle Eastern countries. Prior to 1970, the practice of EIA was less identifiable and occurred in an ad hoc manner. O'Riordan and Sewell (1981, p. 10) note the leadership of the geographer Harlan H. Barrows who, in the late 1930s, conducted an investigation into the environmental and social effects of the Grand Coulee Dam on the Columbia River in the State of Washington. As a geographer Harlan had the expertise necessary to combine social and biophysical studies of the environment into a more holistic view of environment impacts.

However, in the 1960s Australian geographers failed to capitalise on the long-standing environmental strengths of the discipline. Instead, they approached environmental education in a fragmented way (Cardew 1997). Consequently, the demand for a broader type of environmental approach was taken up by new academic programs in environmental studies, at the expense of geography as a discipline, and these started to become important in the 1970s. Since then, many other disciplines have introduced environmental study programs. There is, for example, environmental engineering, environmental law, environmental medicine, and environmental politics, to name a few. It is clear that many disciplines have reacted to public demand for a strong environmental perspective.

Geography is particularly well placed to provide a holistic environmental approach because it spans both the social and physical sciences. However, only those geographers who have the ability to use this background expertise to go beyond their own disciplinary boundaries and conduct applied environmental studies in an interdisciplinary manner can capitalise on this advantage. For example, some projects may require archaeological, ecological, hydrological, meteorological, or socio-economic and cultural investigations, all of which are strongly linked to aspects of geographical enquiry. However, many environmental impact studies require investigation of planning legislation and other environmental laws; assessment of engineering structures; and interpretation of economic techniques such as contingent valuation. Notwithstanding the broad range of geographical enquiry, many of these investigations will require a different disciplinary approach.

Although the 1970s was the turning point in the development of EIA, the previous 20 years had seen a growing concern about the impact of development on the environment. This concern arose partly from the failure of existing economic techniques (such as cost–benefit analysis) to address the long-term environmental and social consequences of development projects. A classic example of this, given by Clark (1983), is the Aswan dam project in Egypt. At the time it was thought that the benefits would outweigh the costs, even though no modern EIA had been carried out. However, deleterious secondary effects—such as the reduced importance of the Mediterranean fishing industry, the loss of agricultural productivity downstream, increased water-borne disease, and reduced water quality upstream—were largely unforeseen impacts, which detracted from the economic benefits. According to Clark, when EIA was first conceived it was considered as a supplement to cost–benefit analysis: early EIA was often used as another technique for collecting information rather than being seen as an integrated process and therefore part of the decision-making.

O'Riordan and Sewell (1981) identify a number of other factors that they suggest were important in leading to the introduction of EIA. The first of these was the growing scale and associated repercussions of resource development schemes, particularly in the post-war period. Second, there was a proliferation of environmental groups that were better informed, had better tactics, and began to use public figures, lobby groups, and even engage in legal arguments (particularly in the United States). Finally, there was a lack of coordination between the following entities: public agencies involved in promoting schemes, private development companies, and public regulation agencies.

The US National Environmental Policy Act

The evolution of EIA is closely tied to the development of the US National Environmental Policy Act (NEPA). The factors mentioned above were instrumental in the evolution of EIA through this Act. These factors and the evolution of the NEPA represent a shift from economic accounting techniques, such as cost–benefit analysis, to a proper consideration of environmental matters. Since 1969, there has been nearly 30 years of litigation involving provisions of this Act. The courts have been interpreting the process, setting precedents, making rulings, and essentially developing guidelines for the evolution of EIA.

An important section of the NEPA is the requirement that any major federal project that could significantly affect the environment must be accompanied by a detailed statement as set out in section 102 of the NEPA. Section 102 of the NEPA provides that:

> (2) all agencies of the Federal Government shall . . .
>
> (c) include in every recommendation or report on proposals for legislation and other major Federal actions significantly affecting the quality of the human environment, a detailed statement by the responsible official on:
>
> (i) the environmental impact of the proposed action,
>
> (ii) any adverse environmental effects which cannot be avoided should the proposal be implemented,
>
> (iii) alternatives to the proposed action,
>
> (iv) the relationship between the local short-term uses of man's [*sic*] environment and the maintenance and enhancement of long-term productivity, and
>
> (v) any irreversible and irretrievable commitments of resources which would be involved in the proposed action should it be implemented

This statement is required not only for federal projects, but also for activities requiring federal permits and federal grants.

There is also a requirement to consult with and obtain comments from federal, state, and local agencies with environmental expertise. This statement is what is now commonly termed the environmental impact statement (EIS), and is explained in more detail later in this chapter. There is a requirement for public disclosure of the EIS, which created the opportunity for debate and for judicial review over environmental matters and compliance with the NEPA. The NEPA also set up a Council on Environmental Quality, which brought out EIS guidelines in 1971 and 1973, and subsequently, in 1978, introduced the important concepts of 'scoping' and 'tiering', which appeared in the regulations of the Act. The concepts of 'scoping' (discussed in Chapter 3 in the Australian context) and 'tiering' are important mechanisms for rationalising the complicated EIS process.

Scoping was intended to narrow the discussions in the statement to issues that are significant and that have not been covered in previous or existing EIS processes. In this manner, it focuses on the issues that are most important for decision-makers and avoids complicated discussion of peripheral issues. Tiering aims to eliminate repetitive discussions from sequential statements on related topics and to avoid matters that have already been dealt with or others that it is too early to deal with. For example, if documents have already been prepared relating to broad policy issues then a subsequent EIS could focus on specific environmental issues and merely incorporate reference to the earlier discussion on policy. As well as alleviating some of the problems with bulky and repetitious EISs, scoping and tiering were also intended to address a major problem with the time and cost of litigation. About 10 per cent of all federal projects for which an EIS was prepared had lawsuits filed, and legal delays could often bring a project to an end. For example, out of 94 EIA cases in 1991, 14 resulted in injunctions (Wood 1995). In the early stages of the use of EIS there was an emphasis on EIS procedures rather than on the content of the EIS. Many agencies became preoccupied with producing a document that would stand up in court rather than one that was useful to decision makers. This focus on process also enabled environmentalists to attack inadequate projects: the EIA process was used to advantage by environmentalists who, in numerous court rulings, refined—and in some ways defined—the role of the EIS. A major legal battle was the Calvert Cliffs ruling. Citizens took action against the Atomic Energy Commission; the construction of a nuclear reactor plant was stopped because the EIS was found to be inadequate (Council on Environmental Quality 1979). Other court decisions demonstrated the value of EIS litigation as a major environmentalist strategy, and this contributed, in part, to the success of the NEPA in environmental protection. However, it is likely that much of the success is due to the timing of the NEPA's introduction: it came into being at a time when there was a rapidly expanding environmental movement and politicisation of environmental protection (Fairfax and Ingram 1981). Public debate, increased interest in environmental law, and a few initial court victories made the environmental movement enthusiastic about the EIS process. In this manner the NEPA assumed a greater importance than it might have otherwise achieved. However, it should be noted that there has been a decrease in the numbers of EISs being produced, with a maximum of around 2000 per annum in the mid 1970s, falling to about 500 per annum in the early 1990s. Wood (1995) attributes this to increased negotiation and to mitigated 'findings of no significant impact' early in the process. The latter finding, based on environmental assessment, means that an EIS will not be produced.

Another factor in the growing support for the NEPA was the proliferation of environmentally oriented law firms that were underwritten by donors such as the Ford Foundation (Fairfax and Ingram 1981). This meant that expertise and funds suddenly became available to environmentalists, enabling them to challenge almost any EIS in the courts. The ease with which EISs could be challenged was due, in part, to the ambiguity in the Act and to the scope it allowed for legal interpretation. Although the NEPA success was related to its timing and its public appeal, it was the interpretation of the NEPA by the courts that has been a major factor in its acceptance.

It is clear that the EIS, introduced through the NEPA, represents a major change in addressing environmental impacts. The EIS embodies an interdisciplinary approach by including environmental, aesthetic, and non-quantified aspects of proposals. It also requires assessment of alternatives and the opportunity to review and comment. Thus, from 1970 in the United States, projects not only required a thorough interdisciplinary investigation, but also required all the project alternatives to be addressed.

Adoption of EIA by other countries

As noted above, it was not long after the introduction of the NEPA that Australia developed its own EIA procedures in 1974. These were developed after the announcement of a Commonwealth (of Australia) EIA policy in 1972. In fact, Wood (1995) suggests that the first non-US jurisdiction to introduce a 'rudimentary environmental impact policy' was probably New South Wales in January 1972, followed by a number of countries that adopted various forms of EIA procedures in the 1970s.

The first decade of EIA and its global diffusion

1969 The US National Environmental Protection Act is introduced and passed into law on 1 January 1970.
1970 California introduces its own EIA legislation.
1972 New South Wales introduces an 'extremely rudimentary environmental impact policy'; Singapore forms a Ministry of the Environment to assess the environmental impact of development proposals; and the Japanese Cabinet adopts EIA procedures for major development projects.
1973 Canada introduces an EIA cabinet directive.
1974 Australia and Columbia introduce specific EIA legislation, and New Zealand institutes EIA procedures by Cabinet minute.
1975 Thailand introduces specific EIA legislation.
1976 France and Ireland introduce specific EIA legislation (although the EIA was not mandatory in Ireland), and West Germany introduces EIA procedures by Cabinet minute.
1977 The Philippines establishes an EIS system by Presidential decree.
1979 The Netherlands introduces a 'government standpoint' on EIA; China's environment protection law (with EIA principles) is adopted by the Fifth National People's Congress; and the Taiwanese Executive Yuan applies EIA principles to various sectors of the economy

Gilpin 1995; Wood 1995

Subsequently EIA has been established throughout the world, although the EIA systems and implementation vary considerably from the mandatory requirements of 'acts, statutes, and regulations' to the more general adherence to 'guidelines' or to more discretionary and ad hoc procedures. However, as Glasson, Therivel, and Chadwick (1994) point out, the various systems themselves (for example, legislation or guidelines) do not necessarily give any real indication of the thoroughness of the EIA. They note that regulations in Brazil and the Philippines are not well carried out or enforced, whereas Japanese guidelines are well implemented, and in the United Kingdom there are some good ad hoc EIAs.

In addition to the varied adoption of EIA in different countries, there are a number of international development and funding agencies that have their own EIA requirements—for example, the United Nations Environment Programme (UNEP), the Organisation for Economic Cooperation and Development (OECD), the United States Agency for International Development, the World Bank, and the Overseas Development Administration (Glasson, Therivel, and Chadwick 1994). Although EIA was introduced to some European countries in the mid 1970s it was a decade later (1985) before there was a European Community directive (85/337/EEC) on EIA, and it was not until 1988 that this was implemented in the United Kingdom with formal EIA legislation (Glasson, Therivel, and Chadwick 1994). In fact Wood (1993) has compared Australian EIA with the British system of EIA, suggesting that the 'old world may learn from the new' and pointing out that the EC's directive on EIA not only has a less comprehensive set of provisions than Australian EIA but also did not become binding on member states until 1988, 14 years later than the Australian legislation.

With the global spread and rapid adoption of EIA, particularly in the early phase, the need to refine and reassess the methods and techniques used in EIA became pressing. This was particularly important given that EIA represented a move away from traditional economic accounting techniques towards a more holistic assessment of potential environmental impacts. While the EIA process involved some adaptation of traditional techniques, there was also an eclectic approach to the use of other techniques at different stages of the EIA process.

METHODS AND TECHNIQUES FOR ASSESSING ENVIRONMENTAL IMPACTS

Much has been written on methods and techniques in EIA (see, for example, Clark, Bisset, and Wathern 1980; Canter 1983; Bisset 1988; Glasson, Therivel, and Chadwick 1994; Wood 1995), including a recent text specifically on methods (Morris and Therivel 1995). It is not intended here to provide any new material on the subject but rather to provide a short overview relevant to Australian practice. Bisset suggests that a distinction should be made between EIA methods that 'have been described alternatively as methodologies, technologies, approaches, manuals, guidelines and even procedures in the literature' and EIA techniques relating to 'predicting future states of specific environmental parameters such as noise levels' (Bisset 1988, p. 47). However, Glasson, Morris, and Therivel (1994) refer mostly to 'major impact identification' and 'prediction methods' rather than techniques, and also include specific quantitative prediction methods in their discussion. In fact, as Wood (1995, p. 160) notes, the terms 'methods' and 'techniques' are often used interchangeably in practice. For this reason, no sharp distinction is drawn between methods and techniques in this text, although the term 'technique' is generally used more for the specific tools of impact identification and prediction, and is generally used in relation to particular issues.

There are no hard and fast rules for which methods or techniques to use in assessing environmental impacts, and the choice of method or technique may relate to different approaches for determining the significance of various impacts—although even the term 'significance' is interpreted in different ways (for example, it can refer to statistical significance, ecological significance, social significance, or even significance to the project (Beanlands and Duinker 1988)). There are also numerous methods and techniques that focus on different stages in the EIA process, such as the baseline description of environmental parameters, impact evaluation, impact prediction, and impact mitigation. While each of these is important, Glasson, Therivel, and Chadwick (1994) point out that the whole exercise of EIA is about prediction of impacts and that this should occur from the early stages—when alternatives are being considered—through to project planning, mitigation, monitoring, and auditing.

One problem is that, even at the early stages of the EIA process, there is a need for a certain level of detail about potential environmental impacts in order to make informed decisions on:

- whether or not the project or action should be subject to the formal EIA process
- which issues are more important for further investigation
- how rigorous these studies should be.

From the proponent's point of view, detailed environmental investigations can be useful if they assist project planning and ultimately save money, but there is often debate over expensive scientific environmental research where there is a lack of baseline data and uncertainty about the predictability of impacts. For this reason, it is important for the scoping process to be as transparent as possible and for it to result in an agreed guidelines document specifying the scope and level of investigations to be undertaken.

Methods and techniques for environmental impact investigations are diverse and have been categorised quite differently by different authors, as shown in Table 1.1. It is interesting to note that Bisset

(1988) discusses simulation modelling as an important development in EIA methods, but six years later Glasson, Therivel, and Chadwick (1994, p. 112) omitted discussion of simulation models because they considered them to be 'still relatively undeveloped and have, to date, only been applied to problems involving a few environmental impacts'. Gilpin (1995), on the other hand, does not attempt such a categorisation of techniques but does include cost–benefit analysis and economic valuation techniques, in his general discussion of EIA methodologies. He has been criticised for the inclusion of these techniques, which according to Brew (1995) are not strictly part of environmental assessment. As noted above, it could be argued that EIA was, in part, developed to compensate for the failure of such methods to properly address environmental issues. There is also a clear overlap between what Thomas (1996) terms evaluation methods and what Glasson, Therivel, and Chadwick (1994) term impact identification methods. In fact the latter authors discuss evaluation methods in a separate chapter from their discussion on impact identification methods, while acknowledging the usefulness of many methods for both purposes.

Notwithstanding the various classifications of EIA methodologies, there are a number of common methods (including specific techniques) that deserve mention. The following discussion provides a brief overview of some of the more important methods, but the reader is encouraged to consult other references for detail on these (Rau and Wooten 1980; Bisset 1983; Bisset 1988; Gilpin 1995; Glasson, Therivel, and Chadwick 1994; Morris and Therivel 1995; Thomas 1996).

Table 1.1 Categorisation of methods and techniques for EIA investigations

Developments in EIA methods	*Impact identification methods*	*Evaluation methods*

• index methods
  • checklists
  • multi-attribute utility theory
• systems diagrams
• simulation modelling
• sound ecological principles
	• checklists
	• matrices
	• quantitative methods
	• networks
	• overlay maps
		• ad hoc methods
		• checklists
		• matrices
		• overlays
		• systems diagrams
		• networks
		• quantitative or index methods
		• mathematical models

Sources: Developments in EIA methods—Bisset 1988; Impact identification methods—Glasson, Therivel, and Chadwick 1994; Evaluation methods—Thomas 1996

Matrices became important in EIA methodology following the development of the Leopold matrix (Leopold et al. 1971), a two dimensional chart with a horizontal list of 100 project actions (such as modification of a regime, land transformation, and construction or resource extraction) and a vertical list of 88 environmental components (such as soils, underground water, erosion, and endangered species of flora or fauna). This matrix is used to identify the magnitude and importance of any potential impact arising from actions associated with a project. Each cell of the matrix is used to score both the magnitude and the importance of the impact out of ten and then the matrix is accompanied by a discussion of the significant impacts in the matrix (the ones with the higher scores). Although the Leopold matrix has been widely used for different types of projects, it has a number of shortcomings: it only focuses on direct impacts; it doesn't encompass interactions between environmental components or between indirect effects; it has the disadvantage of being cumbersome for specific projects; the scoring technique is likely to be subjective; and there is a danger in deriving composite scores from such a technique.

There are other forms of matrices, such as:

- weighted matrices
- interaction matrices
- stepped matrices
- time-dependent matrices
- three dimensional matrices.

These more complex matrices attempt to address some of the problems of the Leopold matrix by incorporating aspects such as a third dimension (for example, socio-economic factors), time dependency, and magnitude of impacts—or by providing a weighting system for different impacts.

Checklists are also a commonly used method in EIA and are popular because of their ease and rapidity of use. In their most basic form, simple checklists (see Table 1.2) identify a series of potential environmental impacts but do not link these to specific project activities. However, there are also more detailed categories, such as:

- questionnaire checklists
- descriptive checklists
- threshold of concern checklists
- scaling checklists
- scaling–weighting checklists.

The questionnaire checklist is the type that has been used for the proposed Australian EIA criteria (see Chapter 8). Descriptive lists may include the type of data required together with suggested techniques for

Table 1.2 Simple checklist of potential environmental effects for a hypothetical marina development

Potential environmental effects	*Effects in the construction phase*	*Effects in the operational phase*
antifouling paints		✓
basin dredging	✓	
disturbance to sea-bird colony	✓	
employment	✓	✓
heritage sites	✓	
tourist pressure on coastal environment		✓
infrastructure		✓
littoral drift/sediment accumulation		✓
loss of seagrass	✓	
maintenance dredging		✓
noise	✓	✓
quarrying of rock for breakwaters	✓	✓
social impact	✓	✓
traffic	✓	✓
vegetation clearance	✓	
water quality	✓	✓

determining water quality, such as turbidity measurements or faecal coliform content. Threshold of concern lists (Table 1.3) merely specify the threshold level at which each impact is considered significant enough to warrant concern. Scaling checklists usually provide a common subjective scaling (for example, –3 to +3) to each parameter in order to determine the direction and magnitude of the impacts. A more complex checklist is the scaling–weighting type, such as the Batelle system (Dee et al. 1973), also known as the environmental evaluation system. This system involves giving each of 78 environmental factors an environmental quality scale. The resultant scales are then weighted to develop a composite environmental score. The figures can then be recalculated based on predicted impacts and altered scales for each of the environmental factors. This can be repeated for different scenarios so that the relative impacts of each scenario are assessed. Although the Batelle system is quite complex, it has been criticised for being inflexible to different types of projects, for being subjective in its summation, and for lacking the identification of interaction between environmental effects.

Another method that has been widely used is the McHarg overlay technique (McHarg 1968), which originated from environmental planning prior to the existence of formal EIA. This technique involves

Table 1.3 Threshold of concern (TOC) checklist for a hypothetical marina development

Environmental effect	*Criteria*	*TOC*	*Impact from marina scheme 1*	*TOC*	*Impact from marina scheme 2*	*TOC*
disturbance to sea-bird colony	population of fairy penguins	300	200	yes	250	yes
tourist pressure on coastal environments	number of boats in nearby coastal lagoon	25	20	no	30	yes
littoral drift/sediment accumlation	extent of sand accumulation at breakwater	50 000 m³	75 000 m³	yes	40 000 m³	no
water quality	secchi disc visibility	1 m	0.5 m	yes	2 m	no
vegetation clearance	extent of clearance	5 ha	6 ha	yes	3 ha	no

the overlaying of various transparent map sheets, each of which contains a different environmental variable. The technique, which should be considered as a rapid first approximation of the spatial distribution of potential impacts, has been criticised because of its simplicity and subjectivity, although as Glasson, Therivel, and Chadwick (1994) note, overlay maps are easy to understand, popular in practice, and an excellent way of showing the spatial distribution of impacts. To a large extent the McHarg technique has been superseded by the modern use of computer-based geographic information system (GIS) data bases and GIS software programs such as ARC/INFO (Figure 1.1). These are capable of examining a multiplicity of layered spatial data and at the same time providing complex mathematical computations of various interactions between environmental variables.

Glasson, Therivel, and Chadwick (1994) identify matrices, checklists, and overlay maps, together with various quantitative methods and networks, as commonly used methods for impact identification, although they acknowledge that they may also be of value in impact prediction. They refer to the multiplicity of prediction methods and point out that they can be classified in a number of ways, such as

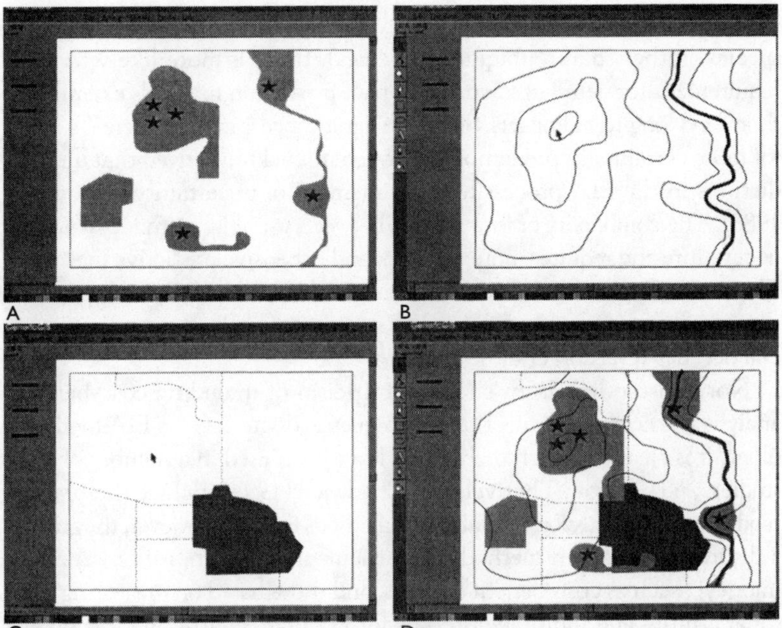

Figure 1.1 Example of a composite, computer-based GIS map (D), incorporating three variables (A, B and C)
Source: Diagram generated by Brett Bryan.

extrapolative or normative, or according to whether they are partial or holistic in their coverage. They also classify methods according to their form, as illustrated by various types of models:

- mechanistic or mathematical models
- mass balance models
- statistical models
- physical, image, or architectural models
- field and laboratory experimental methods
- analogue models (Glasson, Therivel, and Chadwick 1994).

Methods for the prediction of environmental impacts must contend with the following issues:

- the level of specificity required
- the complexity of interactions between variables
- the magnitude of the impact
- the different spatial and temporal scales involved.

The level of specificity required will, in general, be greater for more significant impacts, but this in turn relies on the method for determining significance. For example, if there is potential impact on an endangered plant species (rather than a ubiquitous species), there is more likely to be a requirement for detailed scientific impact prediction methods examining short- and long-term impacts. Since the significance is greater, there is a need for more certainty in prediction, although it is acknowledged that all predictions in the EIA process have an element of uncertainty (De Jongh 1988). The complexity of interactions between variables introduces further uncertainty and requires more complex and expensive predictive methods. There is also a problem with long time-scales, particularly where social impact is concerned and where project impacts are likely to extend over several decades (Glasson et al. 1987).

Notwithstanding Brew's (1995) criticism of including cost–benefit analysis and economic valuation techniques in discussions on EIA methodology, it is clear that such techniques have been used in a number of EIA studies, and Glasson, Therivel, and Chadwick (1994) include these techniques in their discussion of evaluation methods in EIA. However, they make a distinction between methods that assume a 'common utilitarian ethic [money]', such as cost–benefit analysis, and those based on measuring personal utilities and multiple criteria—for example, multi-criteria analysis, decision analysis, and goals achievement. The reader is referred to their discussion on these methods for further information. It is worth noting that a number of environmental monetary valuation techniques, such as contingent valuation, the travel–cost approach, and hedonic pricing techniques (Gilpin 1995), have been devised in an attempt to place monetary value on environmental parameters.

However, in many cases it is not possible to reduce environmental values, such as aesthetics or desire for preservation, to monetary values. For example, the now defunct Australian Resource Assessment Commission was strongly criticised by conservation groups when it used the contingent valuation method in its 1991 Kakadu Inquiry—the commission asked people to place a value on retaining an area for conservation. This criticism relates largely to philosophical differences about the validity of the technique itself. More recently, in another attempt to place monetary values on potential environmental impacts, Australian research is being conducted into a new 'choice modelling technique' (Morrison et al. 1996).

While prediction methods are critical from the early stages of planning through to auditing, Wood (1995, p. 212) suggests that 'the mitigation of environmental impacts is the principal aim of the EIA process'. For this reason mitigation methods are an integral part of environmental impact investigations. In fact, mitigation of potential impacts is likely to occur through the various stages of the process so that early plans and alternatives are modified to avoid impacts, and further investigations may lead to other mitigation strategies. Mitigation measures vary according to the scale and level of impact and have been classified by the UK Department of Environment into three categories: avoidance, reduction, and remedy. For example, the only mitigation for a project resulting in the loss of a pristine and unique wilderness area may be to avoid the impact by adopting the 'no-project' alternative. Impacts from other projects may be made more manageable by reducing the impacts themselves (see Table 1.4).

In other cases mitigation measures may require a remedy in the form of financial or environmental compensation. For example, a financial bond or bank guarantee may be required where there is risk of significant environmental impact and where certainty in the implementation of the mitigation strategies is required. Alternatively, there may be a trade-off between the loss of one nature reserve to development and the provision of a new one nearby.

As noted at the beginning of this section, there are no rules for which methods to use in environmental impact investigations, but it is likely that there will be a variety of methods used at different stages of the process—from the preliminary planning, through the consideration of alternatives, baseline studies, prediction of impacts, evaluation, and mitigation, to monitoring and auditing. Although some of the methods

Table 1.4 Example of selected mitigation for a small mining operation

Impact	*Mitigation method*
noise from trucks	restrict working hours
excessive dust	spray roads with water
contamination of underground water	preclude mining below watertable
visual impact	stage the mining operation and use fast growing trees/shrubs for screening
slurry entering natural stream	install silt dams
dangerous access	relocate access for adequate line of sight from other vehicles
native vegetation clearance	preserve topsoil and seeds, and implement a staged re-vegetation program
cumulative impact (combined with similar small mining operations in the area)	implement a broader strategic environmental assessment

have evolved for specific purposes, many of the methods have relevance for a number of stages in the EIA process.

Common elements of current EIA procedures

It should be pointed out that EIA as a 'process' is sometimes confused with the term 'EIS', which is a document or a series of documents. There are also other types of a documents or procedures that are part of the EIA process. For example, there may be a regulation requiring a government officer to advise on environmental matters prior to a decision on a routine development application. At the other end of the scale, there may be a major government inquiry involving extensive public consultation and numerous environmental investigations and subsequent reports that, collectively, are far more complex than a standard EIS. The important aspect of these elements of EIA is that each of them is part of a process that examines the environmental consequences of actions, in advance. Most importantly, the results of these investigations are used as an aid to more informed decision-making.

EIA is not just another data collection tool. It is used for the sound assessment of proposals (including proposed policies, programs, plans, actions, and projects). It moves away from purely technical, economic, and political criteria and includes environmental, health, and social impacts. There are a number of fairly standard steps in the EIA process and these are summarised on page 19. The early stages relate to 'screening' and 'scoping', which are part of the US terminology for EIA and are discussed further in Chapter 3. The first step in an EIA is to 'screen' which proposals might have significant environmental impacts and should therefore be subject to formal EIA procedures. Once that decision has been made, it is then possible to 'scope' (or focus) on the potential impacts and the significant issues that are likely to be important in decision-making.

The next step is to examine what is being proposed together with any alternatives and the 'do nothing' option. There needs to be a clear and detailed description of the proposal so that specific potential impacts can be identified and environmental investigations can be undertaken. In some project-based EIAs the proposal is site-specific and financially driven, making alternative project locations less relevant for decision makers. However, a consideration of alternative locations, operations, processes, and the scale of the proposal should be an important consideration in the EIA process. Having defined these parameters, there needs to be an accurate description of the existing environment as well as prediction of what any future environmental changes might be if the site were not subject to human impact. It is then possible to identify key environmental impacts that might arise if the proposal were to be implemented. This should include the direction and scale of the predicted impacts together with their short- or long-term nature. Once the potential impacts have been identified, it is necessary to determine which ones are more significant and to examine mitigation measures designed to ameliorate those impacts. If the impacts have critical threshold levels—that is, if they reach a threshold that is critical—it may be necessary to modify the proposal, use alternative procedures, or provide some form of compensation.

Standard steps in the EIA process

- screening (of projects that require EIA) and scoping (focusing on the key significant issues)
- consideration of alternative proposals and description of the proposed actions
- description of the environmental baseline
- identification and prediction of key impacts
- evaluation of the significance of impacts and proposal of mitigation measures
- EIA presentation, and public consultation and participation
- EIA review and decision-making
- post-decision monitoring and auditing

An important step in the EIA process is public consultation and participation. The public can become involved in various stages of the EIA process, but the most common stage is after the documentation on the proposal and its potential impacts has been prepared and publicly released. Given that it is publicly available, it is also important to produce a clear, readable, and well-presented document that does not have a technical focus (detailed technical material is usually attached as appendices). Public consultation introduces an element of accountability to the process. But it is important that the public be involved in a meaningful manner: appropriate consideration needs to be given to the issues in the consultation process, and feedback should be provided on the extent to which these issues have been considered in decision-making.

After the public exhibition of EIA documents, it is up to the assessing authority (usually this is the government but sometimes it is an independent authority) to review the documents and make recommendations for decision makers. The review of the document involves assessing its quality, and this includes consideration of the degree to which it has considered both public comment and the expert opinion obtained by consultation following each stage of documentation. Once the assessing authority has made its recommendations, it is up to the decision makers to consider them when making their decision. The weight given to the EIA varies considerably depending on the jurisdiction and the particular proposal.

Finally, if the proposal associated with the EIA is approved, there is a need for monitoring, which may be part of the approval conditions. Another important but less-used step in the EIA process is an audit comparing actual and predicted environmental impacts. While monitoring and auditing are useful tools for assessing the effectiveness of the EIA as part of the decision-making process, it is also important to provide for amelioration measures should any critical threshold levels or unacceptable impacts be identified.

While the application of EIA can occur at different levels, it usually focuses on projects rather than policies, programs or policies. Generally, EIA results in a formal document that, although the nomenclature varies around the world (including Australia), is most commonly referred to as an environmental impact statement (EIS), for which there is a general consensus on content (see the boxed material below). There are a number of other names for EIA documents, some of which are virtually synonymous with the EIS—for example, the environmental statement (ES) in the United Kingdom, and the environmental impact report (EIR) in Canada. In Australia there are also other terms for what are essentially EISs—for example, environmental effects statement (EES) in Victoria, and environmental review and management programme (ERMP) in Western Australia. Other Australian documents, such as the public environmental report (PER), provide less detail and are designed for projects with a reduced scale of environmental impact.

Standard content of an EIS

- summary
- justification or rationale for proposed action
- alternatives to the proposed action
- description of the environment that is likely to be affected
- predicted environmental consequences of the proposed action
- list of authors of the report

It is the importance of the EIS in the application of EIA that causes some confusion between the two terms. This may be because the EIS is locked into a very formal and visible process compared with many other EIA procedures. Although it would theoretically be possible to discuss the application of all levels of EIA within a system, including ad hoc and less formal EIA procedures, it is much easier to examine the formal EIS process, which is a public process involving published documents and is readily comparable with procedures elsewhere.

An EIS usually contains either an executive or non-technical summary at the start of the document to facilitate its readability for non-expert groups such as the general public or politicians. The document then outlines the need for the proposal and identifies any alternatives. The main section of the EIS usually comprises

- the baseline data on the environment likely to be affected
- the predicted impacts of the proposal on that environment.

Finally the validity of the report and its predictions are backed up by a list of references and a list of the authors of the report. Australian examples of EISs are discussed in Chapters 6 and 7.

The dominance of the EIS in the EIA process has led many developers to respect and even fear it, and it has been the butt of many jokes and cartoons. The EIS is often made the scapegoat for project delays, additional costs, and public opposition, but in reality environmental investigations and public consultation should already be an integral part of a well-designed project and can often save money rather than create financial difficulties.

After more than 25 years of formal EIA requirements and the dominance of the EIS, there have been numerous reviews, particularly within specific jurisdictions, of the role of EIA and of the usefulness of its function. As Wood (1995) notes, it is generally accepted that at the US federal level EIA has been effective in mitigating environmental impacts and in changing the attitudes of decision makers. Elsewhere, it is sometimes mistakenly assumed that the introduction of EIA procedures alone is sufficient to change the attitude of decision makers. While it is difficult to devise an objective evaluative framework, it is clearly important to know if EIA is effective in influencing decision-making, in making accurate prediction of impacts, in assisting in the achievement of management objectives, and in producing cost-effective studies consistent with proposed time-frames. For this reason, recent interest has been shown in the study of the effectiveness and efficiency of EIA, with important work being done in Canada (Gibson 1993) and also within a broader international framework (Sadler 1996).

Another important aspect of the EIA process that has received recent attention is the role of EIA as an instrument for achieving sustainable development. Glasson, Therivel, and Chadwick (1994) note the importance of EIA in this regard but emphasise the need to extend beyond project-based EIA to incorporate the environmental impacts at the broader policy level. This requires a strategic approach to defining goals for sustainable development (see Chapter 8). In Australia, sustainable development was considered in developing the national approach to EIA (Chapter 8), but it is only recently that the concept has been incorporated into EIA legislation or regulations (New South Wales and Tasmania).

STRUCTURE OF THIS BOOK

Chapters 2, 3, 4, and 5 provide a background to the existing mechanisms of the EIA process in Australia. Chapter 2 outlines some of the differences in the EIA legislation and administration between each of the Australian states and territories. The following three chapters (3, 4, and 5) explore how these differences are reflected in the standard steps of the EIA process as outlined on page 20. In particular, these three chapters focus on the following processes:

- screening
- scoping
- EIS preparation and presentation
- public consultation and involvement
- assessment or review
- decision-making
- monitoring
- auditing
- appeal rights.

Chapters 3, 4, and 5 do not address the detailed methods and techniques used in individual EIA studies because the methods and techniques vary greatly according to variables such as the specific project location and its unique environmental impacts; the approach and methodology adopted by the EIA consultants; the budget and time-framework for the studies; and variations in political and public involvement in each project. The project-specific nature of many of these variables reduces the usefulness of discussing them according to state or territory boundaries.

Having discussed the theoretical framework of the EIA process and the different mechanisms for its operation in the Australian states and territories, it is necessary to examine how it operates in practice. A simple reading of EIA legislation and associated regulations will provide a good idea of how the EIA process should operate in a particular jurisdiction but may not be a true reflection of what happens in practice. For example, there are often ad hoc procedures, accepted bureaucratic practices, or strategies for public involvement, none of which are written down. For this reason Chapter 6 uses case studies from various states and territories to provide some examples of what actually happens in practice.

The EIA process at the Commonwealth level is discussed in a separate chapter (Chapter 7) since there can be jurisdictional overlaps between Commonwealth, state, and territory EIA legislation. As with the previous chapters, a real example is used to illustrate how the Commonwealth EIA legislation operates in practice.

Given the potential for overlapping jurisdictions in Australia, there have recently been attempts to minimise duplication through agreements on Australia-wide common criteria and principles used in the EIA process. These are discussed in Chapter 8 within the framework of broader environmental initiatives. In particular, Chapter 8 examines the implications—for directions of Australian EIA during the next century—of a growing interest in strategic environmental assessment.

2

EIA legislation and administration in Australian states and territories

This chapter provides a brief comparative review of the legislative provisions for EIA in each of the Australian states and territories. There are a number of differences between the various pieces of legislation containing EIA provisions. In particular, some states, such as New South Wales, South Australia, the Australian Capital Territory, and Queensland (to a degree), have their EIA legislation linked to planning legislation. With the introduction of the new *Environmental Management and Pollution Control Act 1994*, Tasmania has also moved towards increased integration with planning. Western Australia is linked more to environmental protection (although it has recently developed stronger links to planning), while the Northern Territory, Victoria, and the Commonwealth have independent pieces of legislation. All processes are centralised under one main Act in each state with the exception of Queensland which has two key Acts. In addition, the Tasmanian EIA requirement under one piece of legislation is linked to another two Acts. Both the amendments to the New South Wales legislation in 1993 and the new Tasmanian legislation introduced in 1994 include provisions addressing sustainable development in the context of the EIA process.

While every attempt has been made in this book to provide current information for each state and territory, the reader should be aware that EIA legislation is constantly changing. In fact, some of the changes to EIA legislation described in this text did not come into effect until January 1996 in Tasmania, and January 1997 in South Australia. Given the changes to various pieces of state and territory EIA legislation in the 1990s, it is unlikely that additional major changes will occur this century. However, a useful method for keeping informed of such changes is to use the Australian EIA Network on the Web at <*http://www.erin.gov.au/portfolio/epg/eianet.html*>.

It should be noted that state and territory EIA legislation uses the term 'responsible Minister' or 'Minister reponsible for EIA'. These expressions refer to the Minister who is in charge of EIA in that state or territory. Commonwealth legislation, however, uses the term 'action Minister', which refers specifically to the Minister who triggers the EIA requirements. State legislation also refers to the 'Governor in Council', who is at the top of the state legislative hierarchy.

AUSTRALIAN CAPITAL TERRITORY

Provisions for EIA in the Australian Capital Territory are integrated into the planning system under Part IV(Environmental Assessment and Inquiries) of the *Land (Planning and Environment) Act 1991* and Regulations (1995 reprint). A 'Guidelines' document for EIA has been produced by the ACT Planning Authority, outlining the process in clear and simple terms for each level of assessment (ACTPA 1992). Figure 2.1, which is based on information in the Act, the Regulations, and the Guidelines, is a flow diagram showing the general procedures for EIA in the Australian Capital Territory. The principal administering authority is the Environment Planning and Assessment (EPA(ACT)) section of the ACT Planning Authority, which is located within the Department of Environment, Land and Planning. However, the primary responsibility for the final assessment rests with the Environment Minister.

NEW SOUTH WALES

The New South Wales EIA provisions, as with the Australian Capital Territory, are integrated into the state's planning system. They are contained in Part IV ('Environmental Planning Controls') and Part V ('Environmental Assessment') of the *Environmental Planning and*

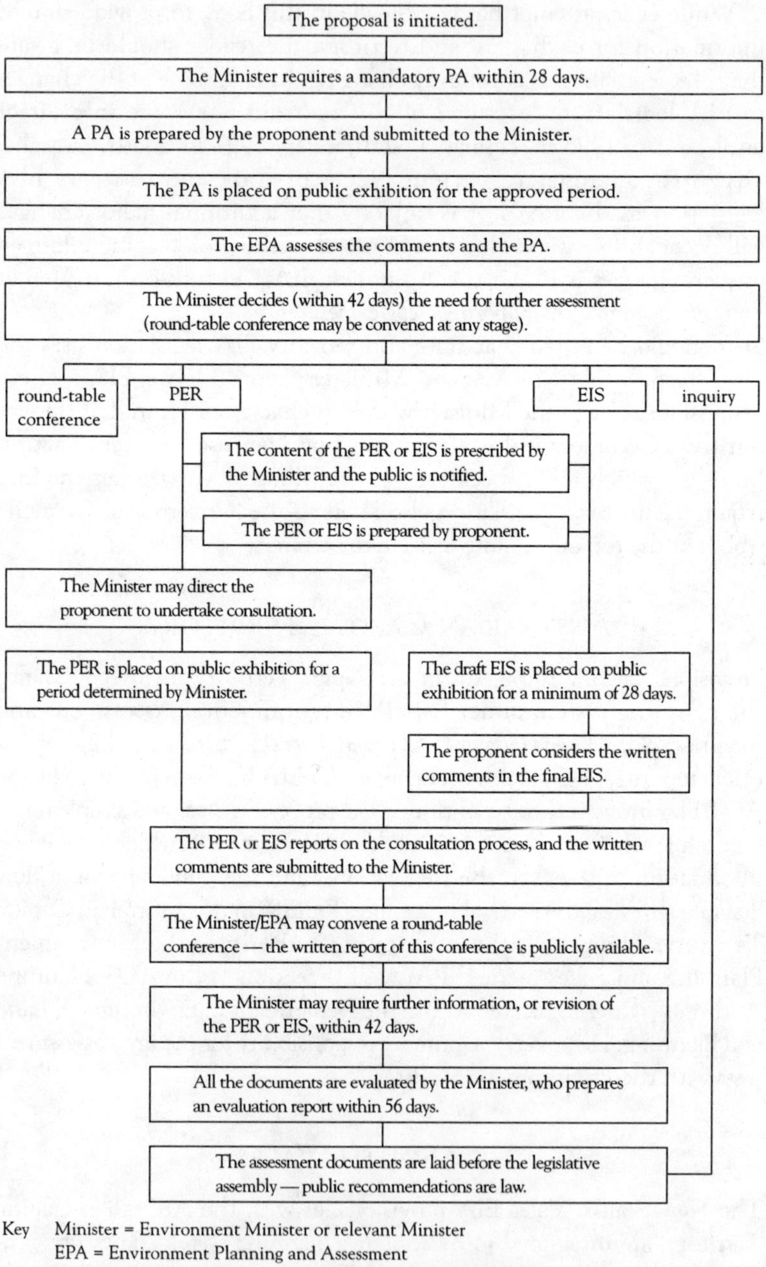

Key Minister = Environment Minister or relevant Minister
EPA = Environment Planning and Assessment
Section of the Australian Capital Territory Planning Authority

Figure 2.1 The EIA process in the Australian Capital Territory

Assessment Act 1979, which was enacted in 1980 with the introduction of the Regulations. There have been several amendments to the Act since 1990, including changes to incorporate Fauna Impact Statements (now called Species Impact Statements) under the *National Parks and Wildlife Act 1974*, and the introduction of 'thresholds' and criteria for designated developments (Cole 1995).

Part V of the Act was amended in 1993, coming into effect on 22 April 1994 via the *Environmental Planning and Assessment Amendment Act 1993* (Department of Planning 1994). The Regulations were also amended at the same time, and clause 5 of Schedule 2 required EISs to include principles of ecologically sustainable development in accordance with the National Strategy on Ecologically Sustainable Development (discussed in Chapter 8). New guidelines for the EIA and ESD process have also been prepared, together with specific guidelines for generic groups of projects (for example, concrete works, bitumen works, cattle feedlots, piggeries, marinas, extractive industries) (ICESD 1996). A consolidated Act and Regulations is available on the Australian EIA Network (Australian EIA Network). The main responsibility for assessment is with the Department of Urban Affairs and Planning via the Director General of the department or the Minister for Environment. A summary of procedures is outlined in Figure 2.2.

NORTHERN TERRITORY

The EIA process in the Northern Territory is generally modelled around the Commonwealth legislation (discussed in Chapter 7), and is administered under the Northern Territory *Environmental Assessment Act 1982*, which formally commenced in 1984 with the introduction of *Administrative Procedures 1984*. Guidelines for the operation of EIA have been produced by the Conservation Commission of the Northern Territory (CCNT 1990). In order to make the Act more consistent with the *Intergovernmental Agreement on the Environment* (DLPE 1995), it was amended in November 1994, and the Procedures were amended in March 1995 and came into effect in May 1995 (DLPE 1995).

The EIA process (summarised in Figure 2.3) is administered by the Minister for Environment (previously the Minister for Conservation), with the Environment Protection Division of the Department of Lands, Planning and Environment maintaining an advisory and coordinating role on the Minister's behalf. Prior to departmental

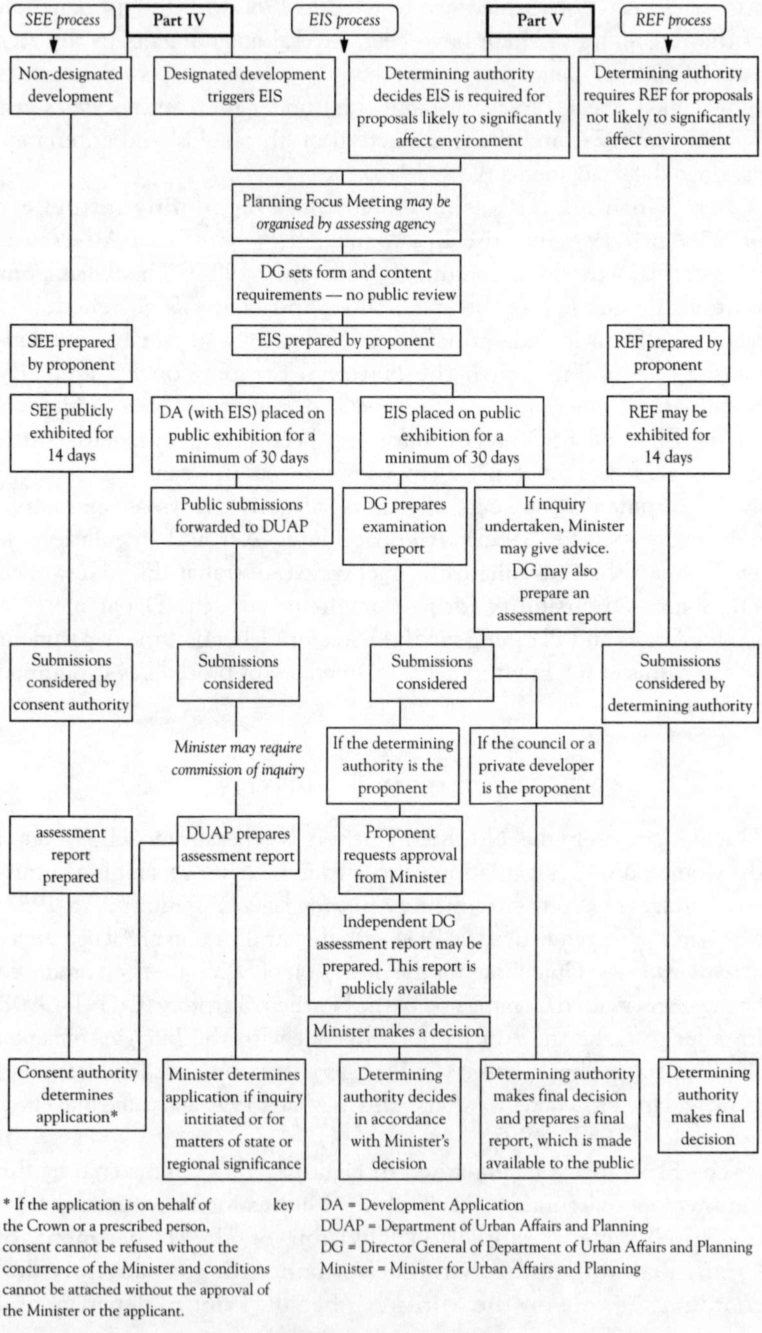

Figure 2.2 The EIA process in New South Wales

restructuring in mid 1995, this function was undertaken by the Conservation Commission.

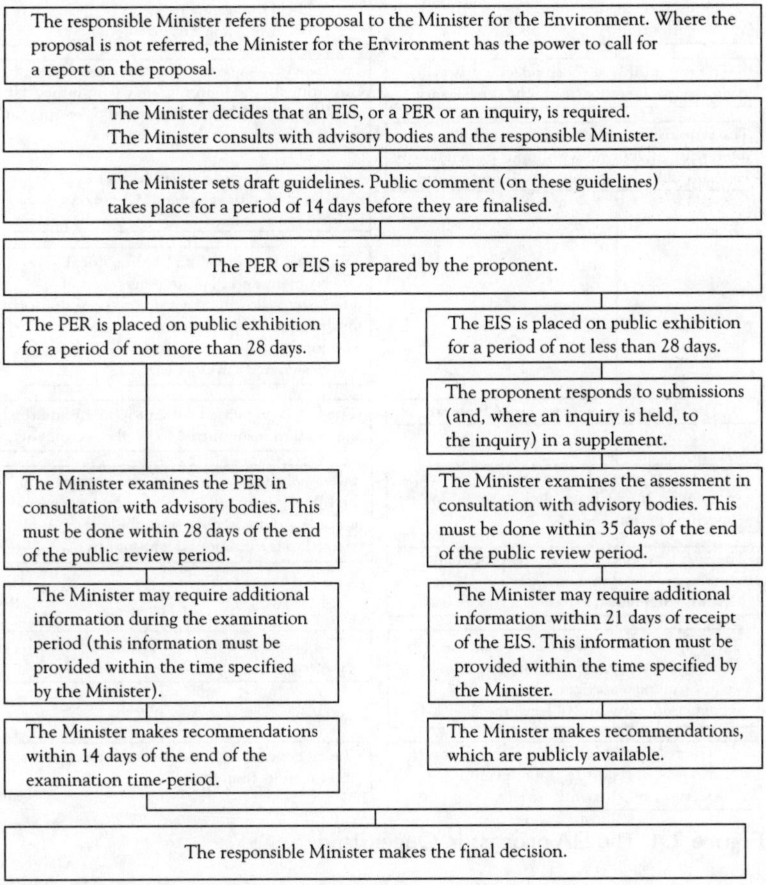

key Minister = Minister for the Environment

Figure 2.3 The EIA process in the Northern Territory

QUEENSLAND

Unlike other jurisdictions, EIA in Queensland (see Figure 2.4 on p. 30) is decentralised, with seven Acts making at least some reference to EIA requirements—for example, the *Integrated Resort Development Act 1987*, the *Canals Act 1958*, the *Clean Waters Act 1971*, and the *Electricity Act 1976* (Coopers & Lybrand 1994). However, the two key statutes and their respective provisions are:

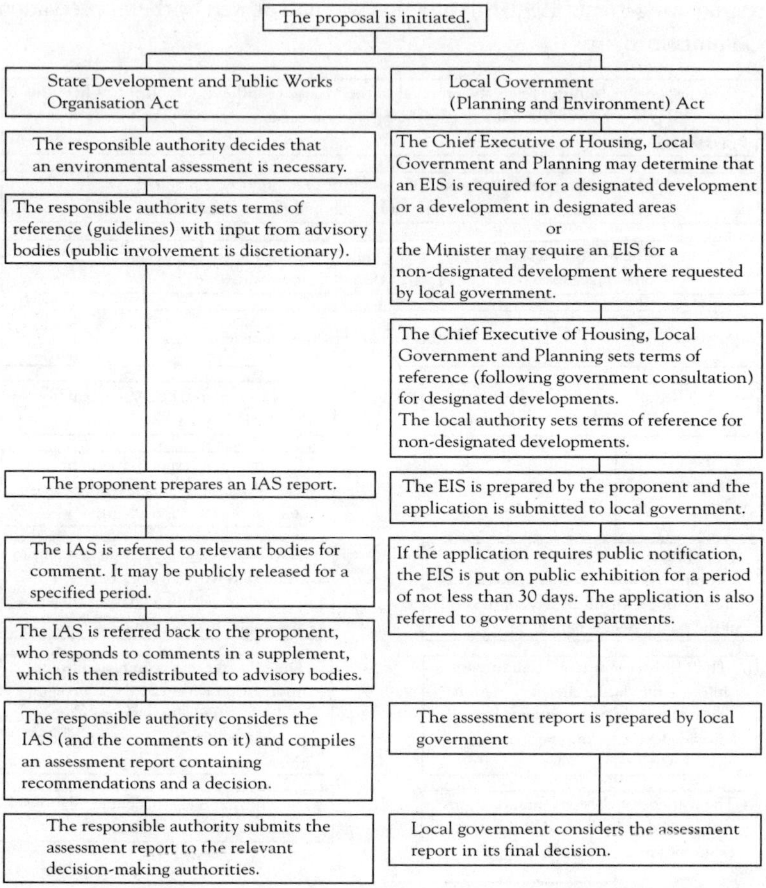

Figure 2.4 The EIA process in Queensland

- section 29, *State Development and Public Works Organisation Act 1971*—this Act is administered by the Co-ordinator General (appointed by the Governor in Council under section 7 of the Act) and this function was delegated to the Director General of the Department of Environment and Heritage in 1988 (Scanlen 1993).
- section 8, *Local Government (Planning and Environment) Act 1990* and Regulations—section 8 is administered by the Impact Assessment Unit in the Department of Housing, Local Government and Planning.

The *Mineral Resources Act 1989* is also a key Act outlining EIA provisions but it is not described in detail here.

The State Development and Public Works Organisation Act does not outline any specific procedures for EIA. However, non-statutory guidelines (the Green Book) have been produced outlining policy and administrative arrangements for EIA at this level (Co-ordinator General, Premier's Department 1987).

SOUTH AUSTRALIA

South Australia has its EIA provisions integrated into the state's planning system through the *Development Act 1993* (and prior to that, the *Planning Act 1982*). The Development Act was amended by the *Development (Major Development Assessment) Amendment Act 1996*, which took effect from 2 January 1997. Division 2 of the Development Act refers specifically to the EIA of major developments or projects (sections 46–8) and has appropriate linkages with the *Environmental Protection Act 1993*. The responsible minister is the Minister for Housing and Urban Development, although there is provision under section 75 of the Development Act for the Minister of Mines to have the same power in calling for an EIA. A set of guidelines has been prepared for the EIA process (see Figure 2.5 on p. 32).

The process is administered by the EIA branch within the Department of Housing and Urban Development, although the initial stages are controlled by the independent Major Developments Panel, which reports to the Minister on the level of assessment required and provides project-specific guidelines after public consultation. In addition, Part 10 of the associated Development Regulations outlines specific provisions and prescribed criteria for the EIA process.

TASMANIA

Prior to 1996, EIA in Tasmania was separate from planning processes, being governed by the *Environment Protection Act 1973* and the *State Policies and Projects Act 1993*. The Acts did not make explicit reference to EIA *per se*, and were guided only by administrative procedures (DELM 1994). However, with the introduction of a new integrated resource management and planning system in 1993, EIA is now governed by the *Environmental Management and Pollution Control Act 1994*, although sections addressing EIA were not enacted until January 1996. Relevant sections of the Environmental Management and Pollution Control Act include Part 3, Division 1 (Environmental Management—Assessment of Environmental Impacts), and Part 5, which outlines the EIA principles.

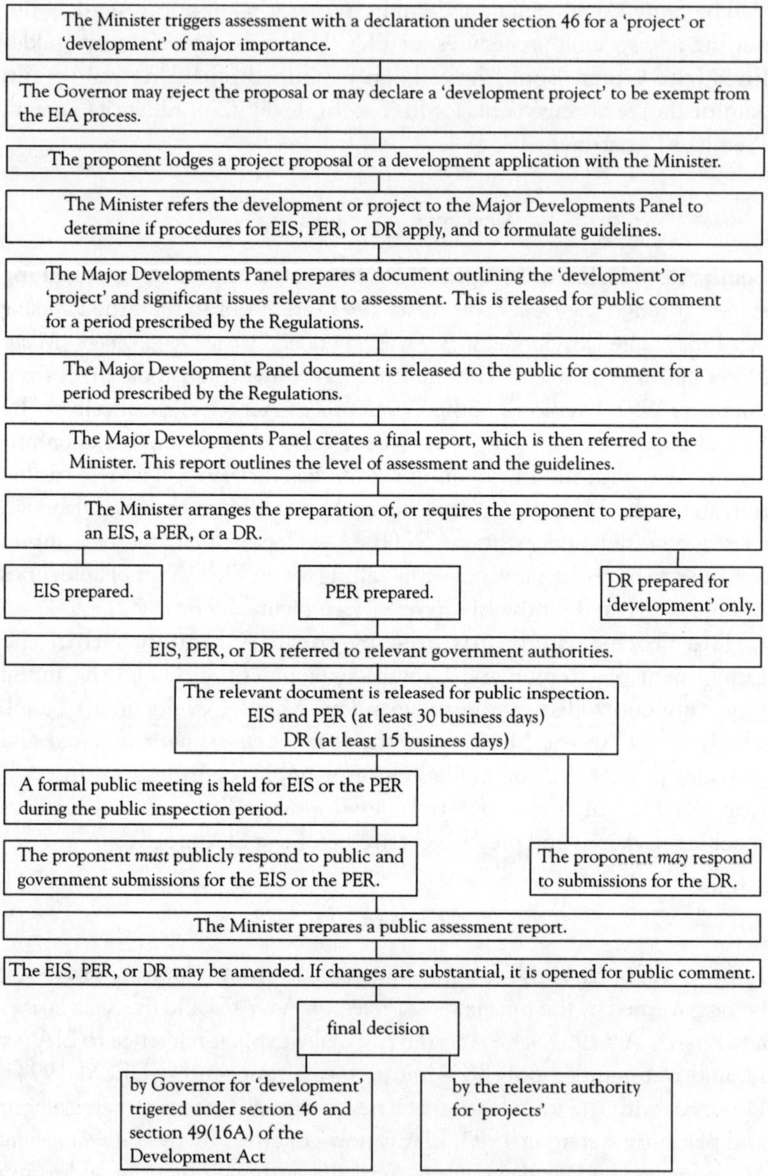

key Minister = Minister for Housing and Urban Development

Figure 2.5 The EIA process in South Australia

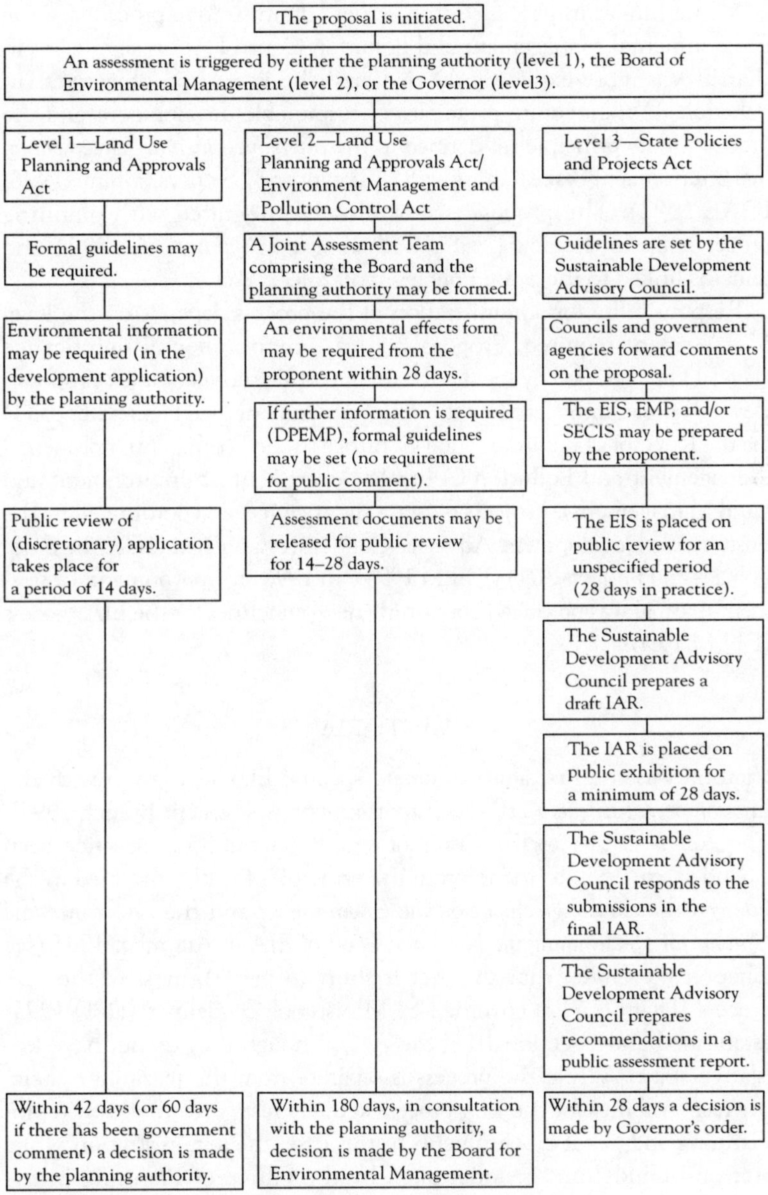

key SECIS = Social, Economic & Community Impact Statement
DPEMP = Development Proposal and Environmental Management Plan

Figure 2.6 The EIA process in Tasmania

While similar in process to the earlier administrative procedures, the Environmental Management and Pollution Control Act is significant in that EIA is more explicitly defined, and the issues and objectives (in Schedule 1) incorporate principles of sustainable development and the precautionary principle (as defined in the *Intergovernmental Agreement on the Environment*—see Chapter 8) (Townsend, S. pers.comm. 1996; IGAE 1992). The process is now closely integrated with planning processes under the new *Land Use Planning and Approvals Act 1993*, and remains linked to the State Policies and Projects Act.

Responsibility for administration of the process depends on the level of assessment required. Proposals of least environmental significance (level 1) are assessed by local government while more complex proposals (level 3) are administered by the state government (see Figure 2.6, p. 33) via the Director of Environmental Control and the Board of Environmental Management and Pollution Control (Department of Environment and Land Management). Projects of State significance are coordinated by the Sustainable Development Advisory Council (established under the State Policies and Projects Act) (DELM 1994). In 1996 an environmental assessment manual was produced containing new guidelines for the EIA process (DELM 1996).

VICTORIA

Victoria was the first state to create specific EIA legislation with the *Environmental Effects Act 1978* (Environment Assessment Branch 1995). The Act is similar to the Commonwealth system, and has since been amended, in 1995, to incorporate the principles for EIA outlined in the *Intergovernmental Agreement on the Environment* and the *Guidelines and Criteria for Determining the Need and Level of EIA in Australia, 1995* (see Chapter 8). Given that the Act is short (4 pages), most of the EIA process (Figure 2.7) is governed by Ministerial Guidelines (DPD 1995) established under section 10 of the Act, although they do not have legislative status. While the process is separate from the planning system, the responsibility for the EIA process actually rests with the Minister for Planning and Local Government. In this case, the administration of the Act and Guidelines is undertaken by the Office of Planning in the Department of Infrastructure (Krohn 1997). De facto assessments may also be triggered under the *Planning and Environment Act 1987* and the *Environment Protection Act 1970*, although the latter is limited to pollution control (Kinhill 1994; Coopers & Lybrand 1994; Environment

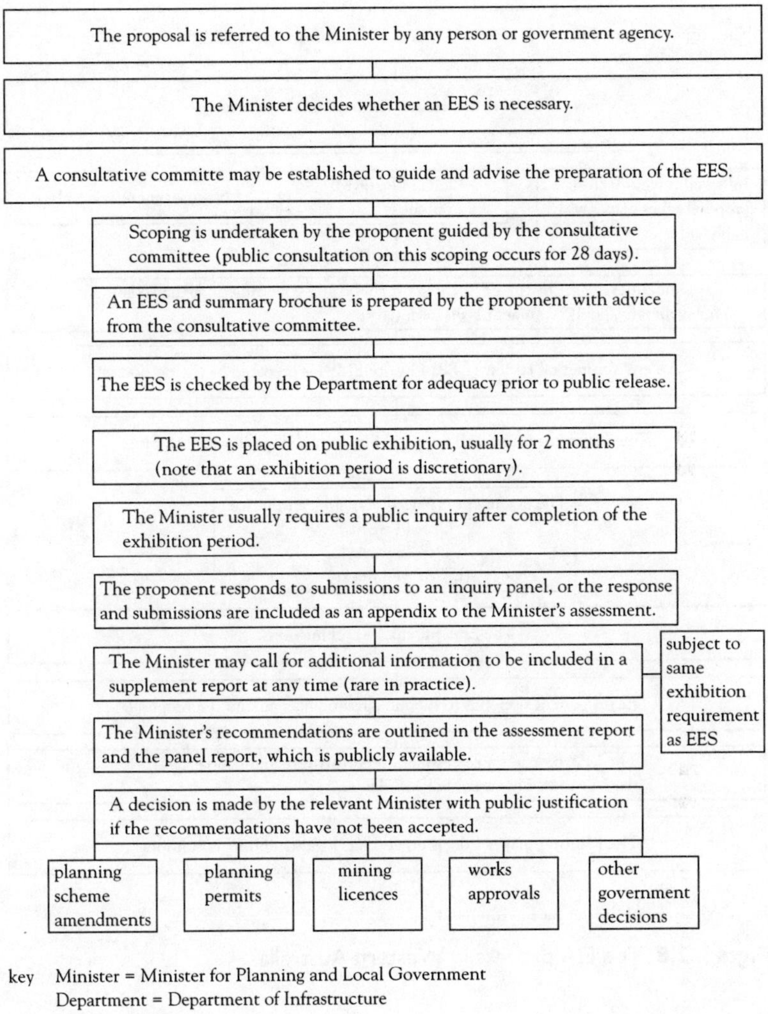

Figure 2.7 The EIA process in Victoria

Institute 1990; Environment Assessment Branch 1995). In both cases, the process of approval, exhibition requirements and/or inquiries (for example, for planning amendment schemes, planning permit requirements, and works approvals) are closely coordinated with the EES process (DPD 1995).

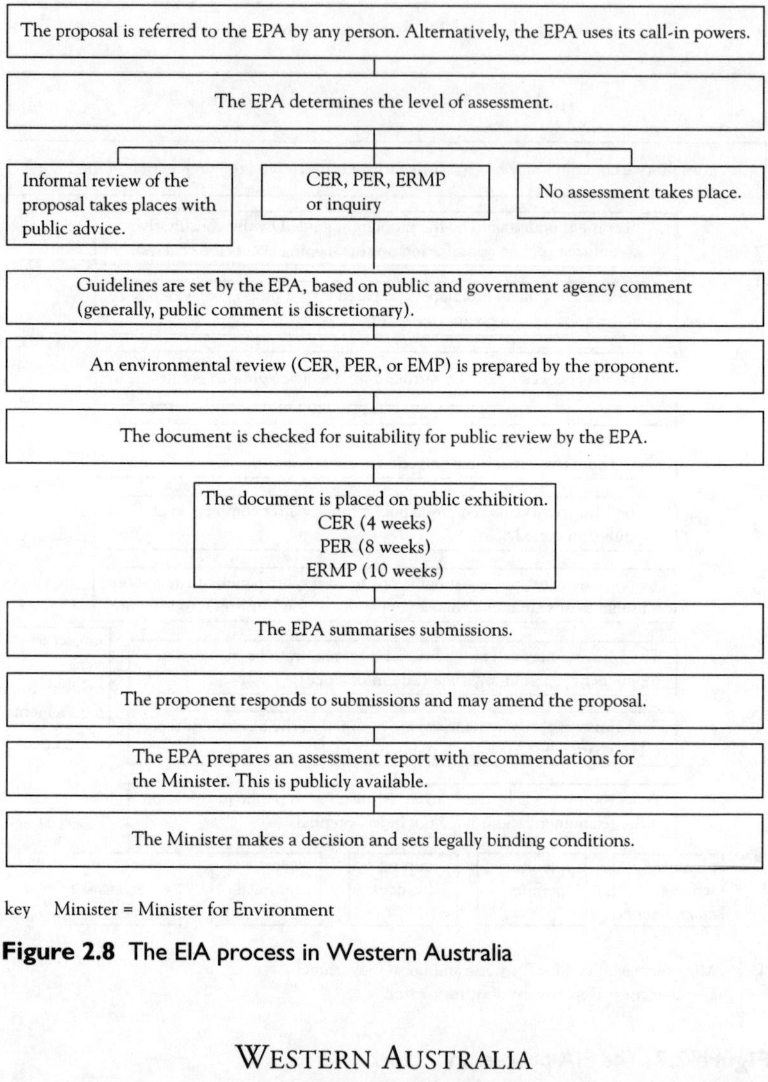

key Minister = Minister for Environment

Figure 2.8 The EIA process in Western Australia

WESTERN AUSTRALIA

EIA processes in Western Australia (Figure 2.8, p. 36) are linked with pollution control legislation, being provided for under Part IV of the *Environmental Protection Act 1986* and the *Administrative Procedures 1993* (established under section 122 of the Act). Unlike EIA in other state and territory jurisdictions, this legislation overrides virtually all other Acts. The Environment Protection Authority (EPA(WA)) retains the primary advisory role, which is largely independent from Ministerial direction, while the Minister for Environment ultimately makes the final environmental decisions (Bailey and English 1991). The independent powers for assessment and decision-making held by the EPA(WA) and the Minister for Environment are unique in Australian EIA processes.

The Department of Environmental Protection manages the overall process for the Authority by gathering information, facilitating briefings and publishing reports on the Authority's behalf (EPA(WA) 1994). Western Australia was also the first state to negotiate joint assessment procedures with the Commonwealth.

CONCLUSION

It is apparent that there are significant differences in both the various pieces of legislation and the administrative arrangements in each of the states and territories. Some of the administrative differences will be discussed in more detail later, but it is worth noting here that there are significant differences between the pieces of EIA legislation that are either linked to planning control, are more independent, or in one case (Western Australia) are linked to environmental protection. It is also noticeable that there have been a number of changes to legislation in the 1990s, and in two of the states amended EIA legislation has only been operative since 1996.

It is difficult to say which is the best system because the various pieces of EIA legislation are linked in different ways to other pieces of state and territory legislation and to administrative arrangements in each state and territory. For this reason, it is easier to achieve agreement between the various states and territories on general principles of EIA and general criteria (discussed in Chapter 8) than it is to attempt to forge a single piece of legislation to apply universally.

In conclusion it is worth noting that the Western Australian EIA system was part of a recent comparative assessment of eight EIA jurisdictions (Wood 1994), including the United States, Canada, the United Kingdom, the Netherlands, and the Australian Commonwealth. This study examined 14 criteria of EIA effectiveness and applied them to each system. Western Australia was the only jurisdiction that satisfied all of the criteria, followed by the Netherlands. Although there is obviously a level of subjectivity in such studies, it does indicate that some Australian EIA legislation is considered highly in global terms.

3

Screening, scoping, and levels of environmental assessment

Prior to any EIA being undertaken it is necessary to conduct a preliminary review (screening) of any proposed projects or actions to determine if they should be subject to the EIA process. This process of screening is essential to rationalise the use of the formal EIA process, thereby avoiding wasting time on proposals that are likely to have less significant environmental effects. The difficult task is to determine which projects are significant in terms of their potential environmental impacts. While this will depend, in part, on the EIA regulations in the relevant jurisdiction, the determination is often made by a discretionary decision as to whether or not the project or action is considered to be environmentally significant. This exercise of discretion usually involves a Minister making a subjective decision that lacks detail in terms of explicit justification or criteria used in reaching the decision. Alternatively, a prescribed list of projects, actions, or criteria is used to trigger the EIA process (an example of a prescribed list is presented in Table 3.1, p. 40). In many cases there is a combination of the two approaches. The public is rarely involved at this stage of the process.

SCREENING OF PROJECTS

Where prescribed lists are used for screening there are some projects that will always have EIA requirements because of their size and potential impacts, and these will automatically appear on the lists. For example, under Annex I of the European Union EIA Directive—a 'Directive' is a framework law outlining requirements that member states of the European Union must implement—a crude-oil refinery automatically has EIA requirements. But requirements for other projects falling under Annex II are less clear (EIA is discretionary for these projects) and depend on a combination of criteria such as the scale of the project and locational sensitivity (Glasson, Therivel, and Chadwick 1994). In Thailand a prescribed list sets out projects or activities requiring EIA reports (Table 3.1) and projects affecting certain areas, such as land reclamation from the sea or construction of roads through national parks, sanctuaries, or declared watershed areas (Tongcumpou and Harvey 1994). In a review of the screening processes for seven different EIA systems, Wood applied his own criteria to each system and concluded that all the systems had adequate screening apart from Australia. He suggests the reason for this is that screening in Australia, at the Commonwealth level, is done by government departments without any real control being exercised by the environment agency (in 1997, the Environmental Assessment Division within the Environmental Protection Group of Environment Australia) (Wood 1995).

Screening and triggering mechanisms in Australian states and territories

In Australia, most screening and triggering mechanisms are discretionary. For example, the screening of projects in South Australia tends to be ad hoc and relies on individuals within government or agencies to identify projects likely to require formal EIA. The EIA requirements under the South Australian Development Act (section 46, as amended 1996) are then triggered if the responsible Minister forms an 'opinion that a declaration under this section is appropriate or necessary for the proper assessment of development or a project of major environmental, social or economic importance'. The difficulty with such a discretionary triggering system is in determining what constitutes a proposal of 'major' environmental, social or, economic importance. Similar discretionary criteria operated under the previous South Australian *Planning Act 1982*, although the Development Regulations (Part 10) associated with the

Table 3.1 Type and size of project or activity requiring EIA reports in Thailand

Type of project or activity	*Size*
dam or reservoir	>10 ML or >15 km^2
irrigation	>12 800 ha
commercial airport	all sizes
hotel/resort facilities in environmentally sensitive areas such as areas adjacent to rivers, coastal lakes, or beaches or in the vicinity of national parts	>80 rooms
mass transit system/expressway	all sizes
mining	all sizes
industrial estate	all sizes
commercial port/harbour	capacity for vessels >500 tons-gross
thermal power plant	capacity >10 MW
industries	
• petrochemical	• using >100 tons per day of raw materials
• oil refinary	• all sizes
• natural gas separation or processing	• all sizes
• chlor-alkaline industry	• tons/day each combined product
• iron and/or steel industry	• ton/day raw materials or combined furnace capacity >5 tons/batch
• cement industry	• all sizes
• smelting industry other than iron or steel	• >50 tons/day production capcity
• pulp industry	• >50 tons/day production capacity
any project or activity developed in the area declared by Cabinet to be watershed class IB	

Sources: Tongcumpou and Harvey 1994, p. 277

new Development Act do contain some guidance as they incorporate criteria based on a draft list of national criteria (see Chapter 8). Elsewhere in Australia discretionary decisions on EIA often depend on determining what is meant by 'environmentally significant' or projects of 'environmental importance'. Prescribed lists of projects, or actions, to which EIA should be applied (or criteria to guide decisions) are not common in Australia, although New South Wales has lists of designated developments and the Australian Capital Territory has a list of defined decisions. An overview of Australian triggering mechanisms for EIA is presented in Table 3.2.

Table 3.2 Triggering mechanisms for EIA in Australian states and territories

State or territory	*Triggering mechanism*	*Responsibility*
ACT	for PAs—prescriptive by defined decisions in Territory Plan (or discretionary requirement by Environment Minister)	relevant Minister
	for PERs/EISs—discretionary	Environment Minister
NSW	Part IV—prescriptive by designated developments (listed in Schedule 3 of the Regulations)	planning authority
	Part V—discretionary for 'environmentally significant' proposals (The guide for this is the guidelines, 'Is an EIS required?')	determining authority (can be action Minister)
NT	discretionary, by referral of proposals from responsible Ministers	Environment Minister
	The Minister for the Environment also has call-in powers.	responsible Minister
Qld	discretionary (State Development and Public Works Organisation Act) but uses thresholds	Local Government Planing and Environment Act—responsible authority
	discretionary (Local Government Planning and Environment Act) referral of designated developments, or development in designated area discretionary non-designated developments approval sought by local government	Local Government Planning and Environment Act —Chief Executive Officer of Housing and Local Government or the Minister for Housing and Local Government
SA	discretionary, may be by referral of major projects of environmental, social, or economic significance	Minister for Housing and Urban Development or Minister for Mines and Energy (advice on level of EIA from Major Development Panel)
Tas	Level 1—discretionary, referred by local government Level 2—discretionary, referred by planning authority or proponent, or by the Director of Environmental Management using call-in powers	Level 1—planning authority/local council Level 2—Board of Environmental Management and Pollution Control and Director of Environemental Management
	Level 3—discretionary, based on criteria of state significance	Level 3— Minister for Environment and Land Management and the Governor
Vic	Discretionary by referral from proponent or decision making authority	Minister for Planning relevant Minister or Governor in Council to resolve any disputes
WA	Discretionary, but broad referral powers for proponent, any other person, Minister for Environment, or EPA, and compulsory referral by relevant decision makers	EPA and Minister for Environment

Australian Capital Territory

Part IV of the Land (Planning and Environment) Act covers both private and public activities, both of which are defined either in the Territory Plan (the main statutory land-use plan) or at the discretion of the Minister for Environment. 'Defined decisions' prescribed by the Territory Plan automatically require a preliminary assessment (PA) of the proposal's impacts, and proponents are notified by the relevant Minister. It should be noted, however, that the Environment Minister can exempt a defined decision from assessment. The Environment Minister may also require a PA for a proposal, and in this case the proponent is notified within 28 days of the decision. If a PA is prepared by the proponent, the document is lodged with the Environment (or relevant) Minister, circulated to relevant authorities, and distributed for public comment for an unspecified period (ACTPA 1992a). Further assessment may then be required by the relevant Minister (or the Environment Minister) within 42 days. Like most EIA processes in Australia, this decision to require further assessment in the form of an EIS or PER is discretionary. Proponents have attempted to avoid triggering such further assessment by preparing large PAs that are almost equivalent to EISs. The key reason for this practice is to avoid the high costs, in time and money, usually required for EIS preparation and process (Norman 1997).

New South Wales

New South Wales has one of the most prescriptive triggering mechanisms in Australia, particularly in relation to Part IV of the Environmental Planning and Assessment Act, which outlines the development application processes for projects (usually private projects) requiring development consent. In this case, an EIS must be required for all 'designated' development applications (section 77(3)(d)). These are listed in State Environment Planning Policies (SEPPs) established under the Act, and in Schedule 3 of the Regulations (which includes actions 'considered to have potential for significant environmental impact and/or where the public interest is served by EIA') (Coopers & Lybrand 1994; Department of Planning 1985).

A major review of this Schedule was undertaken in 1993/94 to cater for changing environmental knowledge, prediction and mitigation technology, community expectations, industry practice, and legislation (Department of Planning 1994). Following extensive consultation, the revised Schedule came into effect on 1 July 1994 with a new emphasis on 'locational triggers', whereby smaller developments may only be triggered in the event that they are near environmentally sensitive locations (Department of Planning

1994). There are also 'call-in powers' for the Minister for Environment to conduct an inquiry (section 119) and to require Ministerial determination for matters of state or regional significance under section 101.

In contrast, Part V of the Environmental Planning and Assessment Act acts as an 'umbrella' (Farrier 1993), protecting the environment from decisions that

- do not fall under the scope of planning controls under Part IV; *and*
- require approval by Ministers and under Acts other than the Environmental Planning and Assessment Act; *and*
- may significantly affect the environment.

In relation to an activity that satisfies the above conditions, if a 'determining authority' (equivalent of a responsible Minister) deems it to be environmentally significant, they may not grant an approval until an EIS is obtained and examined (section 112). In contrast to Part IV, this decision is highly discretionary (particularly as the determining authority is often the proponent). There are, however, new guidelines, 'Is an EIS Required', produced in 1995 to introduce an element of consistency and accountability into this decision (Stone 1997).

Northern Territory

The Environmental Assessment Act applies to both private and public works, and aims to 'ensure, to the greatest extent practicable, that each matter affecting the environment which is, in the opinion of the Minister, a matter which could reasonably be considered to be capable of having a significant effect on the environment, is fully examined and taken into account' (section 4). Although the Act is essentially project-based in practice (Coopers & Lybrand 1994), the word 'matters' could be open to wide interpretation as it is broadly defined in the following way: the formulation of proposals; works and other projects; negotiation, operation and enforcement of agreements and arrangements (at all levels of government); the making of, or the participation in the making of, decisions and recommendations; and the incurring of expenditure.

There are two mechanisms whereby proposals are referred to those responsible for deciding on the need for EIA, both of which are discretionary. First, paragraph 6 of the Administrative Procedures (subordinate legislation enacted under the Environmental Assessment Act) require all responsible Ministers to refer a proposed action under their jurisdiction to the Minister for Environment. The Minister may then require additional information to make a determination on the need for assessment. Second, in the event that a proposal fails to be referred, the Minister for the

Environment has powers to call for a report on a proposed action. Criteria and examples of proposals that may require referral are listed in the Guidelines to maintain some level of certainty, although these have no legislative status. There is also an inter-departmental agreement with the Department of Mines and Energy whereby some matters need not be referred to the Minister; for example, exploration tenements on non-reserve land (Conservation Commission of the Northern Territory 1990).

Queensland

In Queensland, impact assessment not only focuses on 'projects' but is also expected to apply to wider activities in accordance with the *Intergovernmental Agreement on the Environment* and the *National Agreement on EIA* (see Chapter 8) (Department of Environment, pers. comm. 1997) The State Development and Public Works Organisation Act 1971 is aimed at both private and public works or development 'when it appears that the undertaking of such development or works is likely to have major environmental effects' (section 29(2)). Triggering of EIA under the State Development and Public Works Organisation Act is discretionary, but based on thresholds, and is exercised by the responsible authority (as set out in section 29). However, the responsible authority is still required to consult with advisory bodies specified in Schedule III of the Green Book (non-statutory guidelines outlining policy and administrative procedures for EIA) when determining the need for an impact assessment. The decision may also be based on an initial advice statement, required under the Green Book, that includes details of the proposed development.

The Queensland Local Government (Planning and Environment) Act encompasses proposals that

- may have a 'deleterious effect on the environment' (section 8.12)
- are defined as 'designated developments' (section 8.2(2))
- are developments in designated areas (Department of Environment, pers. comm. 1997).

Designated developments are listed in either the local government's planning policy, or Schedule 1 of the Regulations. The latter were amended in 1994 to rationalise the system (for example, smaller projects or less environmentally significant projects would no longer be defined as 'designated developments'). This amendment effectively reduced the numbers of impact assessments from 250 to 100 per year (DPETD 1995). A similar rationalisation also occurred in New South Wales in 1994.

There are two major triggering mechanisms under the Local Government (Planning and Environment) Act, both of which are discretionary. First, all applicants to a local authority must refer designated developments via a request—to the Chief Executive of the Department of Housing and Local Government—for terms of reference. The Chief Executive then determines the need for an EIS, within 20 working days of the referral, following consultation with relevant agencies and local government. Second, the Environment Minister may require an EIS for non-designated developments if local government has sought approval from the Minister or from the Governor in Council (section 8.2(13)). In the past, all designated developments automatically required an EIS to accompany a development application, a more prescriptive triggering mechanism, similar to that established in New South Wales. However, an amendment in 1991 enabled a higher degree of discretion.

South Australia

In South Australia the requirement for an EIA in relation to a 'development' or 'project' is triggered by the relevant Minister using discretionary criteria under the Development Act (section 46(1)). These requirements can be applied to both private and government projects. The Minister responsible for the Development Act is the Minister for Housing and Urban Development, although EIA provisions for prescribed or specified types of mining operations can also be triggered by the Minister for Mines and Energy under section 75 of the Act. Development has a defined meaning under the Act, but the additional use of the term 'project' encompasses activities outside that definition—including, possibly, such matters as policy formulation. The critical decision for the Minister is to decide whether or not the development or project is of major environmental, social, or economic importance. There are no guidelines for determining what is considered to be of 'major', as opposed to 'minor', importance in regard to the Act.

The EIA requirements can be triggered for a specific development or project, for types of development or projects in a defined area, and for general development in a defined area. Once the Minister has made a decision that section 46 of the Act applies, the relevant development application or project proposal must be referred to the Major Developments Panel, whose job it is to determine the level of environmental assessment required using the criteria in the *Guidelines and Criteria for Determining the Need and Level of EIA in Australia* (see Chapter 8).

Tasmania

Under the Tasmanian Environmental Management and Pollution Control Act, the triggering process is not restricted to projects. It encapsulates both public- and private-sector activities, and varies according to three levels of activity defined in section 3. These reflect the degree of environmental significance of the project. The decision to require an EIA is discretionary in each case.

Level 1 activities are those

- requiring a permit under the Land Use Planning and Approvals Act; *and*
- with the potential to cause 'environmental harm'.

At this level, the identification of 'environmental harm' is determined by local government. The phrase 'environmental harm' is defined in section 5 of the Environmental Management and Pollution Control Act as 'any adverse effect on the environment and includes an environmental nuisance'. Level 1 activities are generally those of least environmental significance, are unlikely to attract significant public interest, and are relatively (in relation to levels 2 and 3) small developments (DELM 1994). The actual process of assessment is outlined under the Land Use Planning and Approvals Act.

Level 2 activities include larger and more complex activities (DELM 1996). These are defined in Schedule 2 of the Environmental Management and Pollution Control Act (a minimum threshold of production is usually stated). In this case, the environmental assessment can be triggered via three mechanisms under sections 24, 25, and 27 of the Act. This encapsulates both level 2 activities requiring a permit ('permissible level 2 activities), and level 2 activities falling outside the planning permit process—for example, transport of hazardous waste (DELM 1996). That is, *all* level 2 activities must be referred to the Board of Environmental Management and Pollution Control (the board) to determine the need for an environmental assessment. For those activities requiring permits, this referral to the board is made by the planning authority. For those activities not requiring a permit, referral to the board is done by the proponent. The Director of Environmental Management (of the Department of Environment and Land Management) also has the power to require referral of a level 1 activity to the board to be assessed as a level 2 activity. Informal joint assessment teams are formed for level 2: the planning authority undertakes a 'planning assessment' under section 57 of the Land Use Planning and Approvals Act (although the normal time frames and the decision-making powers of the planning authorities do not apply), while the Board for Environmental Management administers the 'environmental assessment' process under the Environmental Management and Pollution Control Act (DELM 1996).

Level 3 activities are triggered under section 26 of the Environmental Management and Pollution Control Act and are defined as projects of state significance as declared (by the Governor, on the Minister's recommendation) under the State Policies and Projects Act. Criteria used for this determination must include at least two of the following:

- significant capital investment
- significant contribution to the economic development of the state
- significant economic impacts
- significant potential contribution to Australia's balance of payments
- significant impact on the environment
- complex technical processes and engineering design
- significant infrastructure requirements (DELM 1994).

The Sustainable Development Advisory Council must then undertake an integrated assessment pursuant to section 20 of the State Policies and Projects Act. Once declared, the provisions of any Acts or approval planning schemes no longer apply until a decision from the assessment process has been made, and for this reason, level 3 can be seen as a fast-track approvals system. Activities that don't fall under levels 1, 2, or 3 may also be referred to the board by the Director of Environmental Management—for example, large marinas, dams, and major roads (DELM 1996).

Victoria

The Environment Effects Act applies to proposals that could 'reasonably be considered to have or to be capable of having a significant effect upon the environment' (section 3(1)). Thus the Act may apply to both private (section 8(1)) and 'public works, any Ministerial decision or action, any decision by a person or body under Victorian law; and any other works which the Minister for Planning may specify' (DPD 1995, p. 2). Krohn notes that the Act originally focused on public works (Krohn 1997). While the EIA process could apply to policies and plans, the practice has been to focus on projects (Wood 1993).

The triggering process is relatively broad and discretionary, and is clearly outlined within the Guidelines: a proponent or decision-making authority may request advice; any Minister may also cause an environmental assessment to be prepared; and any person may refer a proposal to the Minister for Planning to determine whether the Act applies (DPD 1995). In the case of public works, if there is disagreement between the relevant Minister and the Minister for Planning as to the applicability of the Act, the final decision may be made by the Governor in Council, who may also exempt specified works or classes of works from the Act by proclamation (a proclamation is a notice of such intention that appears in the *Government Gazette*). Similar exemptions also occur in the Commonwealth and the ACT EIA processes.

Once referred, informal discussions with the proponent, with relevant government agencies, with community groups, and with the Department of Planning are initiated to examine the issues and the need for an environmental assessment. If the proponent can demonstrate that environmental effects are already considered sufficiently under the Planning and Environment Act or the Environment Protection Act, an EIA is generally not required (Environment Assessment Branch 1995). The final decision to require an EIA is made by the Minister for Planning, using the criteria in the *Guidelines and Criteria for Determining the Need and Level of EIA in Australia* (see Chapter 8) within 28 days of the request for advice. Importantly, the practice of consultation at such an early stage, particularly consultation with community groups, appears to be a relatively unique and successful feature of the process (Saunders, R., pers. comm. 1996).

Western Australia

Under section 38 of the Environmental Protection Act, all proposals that 'appear . . . likely, if implemented, to have a significant effect on the environment' must be referred to the EPA(WA) via a notice of intent. A 'proposal' is broadly defined as any 'project, plan, programme, policy, operation, undertaking or development or change in land use, or amendment of any of the foregoing' (section 3). Triggering processes are similarly broad. There are four triggering mechanisms:

- compulsory referral from relevant decision-makers
- referral from either the proponent or 'any other person'
- referral from the Minister for Environment (in the event of public concern)
- referral via the EPA(WA) itself—which has the power to require any decision maker to refer a proposal.

As stated by Bailey and English (1991, p. 194), 'under this system it is virtually impossible for politically-sensitive proposals to avoid the EIA processes, a problem which can arise in jurisdictions where the trigger for EIA lies under the control of the "Action Minister"' (for example, under Commonwealth legislation). Section 46 of the Act also provides for amendments of conditions. In this case, if a proposal that has already been assessed changes significantly, it may be referred to the EPA(WA) as a 'revised proposal' and be re-assessed.

SCOPING AND LEVELS OF ASSESSMENT

Once a decision to use the formal EIA process has been made, it is necessary to focus on the important issues. As noted in Chapter 1, the US National Environment Protection Act introduced the concept of scoping in 1978 to narrow the discussions in EIA statements to the significant issues and to avoid some of the extensive, and at times peripheral, EIA documentation that occurred in the early 1970s. As with the screening process, the main difficulty in focusing on issues is in identifying criteria for determining which issues are more significant. Various methods for determining significance are used in both screening and scoping (Tomlinson 1984; Beanlands 1988) but a key difference is that, unlike screening procedures, the public and the proponent are often involved in the scoping process.

Ideally, the scoping process should start with the identification of individuals, communities, local authorities, and statutory consultees likely to be affected by the project. These parties should then be brought together in a meeting with the developer (Glasson, Therivel, and Chadwick 1994, p. 76). In reality this does not always happen and formal scoping is not always included as a stage in the EIA process. In Wood's review of seven different EIA systems, he concludes that scoping is

- a formal requirement for EIA reports under the national US legislation and in the Californian EIA process
- a general requirement in the legislation of the Netherlands, Canada, and Australia
- strongly encouraged in New Zealand legislation, *but*
- lacking in the UK legal provisions (Wood 1995).

Although the UK system provides advice that scoping should occur, the lack of early discussion in the process has been identified as a major limitation of the effectiveness of EIA in that country.

An important part of scoping is the provision of some form of documentation—usually in the form of guidelines—that has been finalised after consultation. This documentation gives the proponent some certainty in terms of what is expected from the EIA investigations. At the same time, the guidelines should be made public so that the proponent is accountable for the standard and the scope of the EIA document.

Levels of assessment vary according to both the strategic approaches to environmental assessment and the specific types of documentation required for projects with different impacts. In relation to strategic approaches to environmental assessment, there are different levels, including the policy stage, the plan stage, and the program stage. Issues dealt with at the broader levels (broader than the specific project level) are not dealt with efficiently at the project stage. In this manner, environmental assessment of policies, plans, programs, and projects can be viewed as a tiered process (Glasson, Therivel, and Chadwick 1994). The tiered nature of the EIA process is further discussed in Chapter 8.

At the project level, there is considerable variation in the terminology used for the different EIA documents, even within Australia. Part of this variation relates to the complex variations in legislation between different states and territories (described in Chapter 2), but much of it relates to different levels of documentation required for projects according to their complexity (and their significance in terms of their potential environmental impacts). For example, the EIS, or its equivalent, is a commonly accepted form of documentation, but in many states the documentation required for less significant proposals can be less detailed than an EIS—for example, the public environment report (PER) under the Australian Capital Territory Land (Planning and Environment) Act. This allows a scoping of issues appropriate to less significant projects and often reduces consultation time, thereby avoiding unnecessary delays for smaller projects. Conversely, to allow for those projects whose potential environmental impacts may have implications extending beyond the local area, many states have provision for some form of environmental inquiry at a higher level than the EIS.

Scoping and levels of impact assessment in Australian states and territories

In Australia, there is a requirement for the Commonwealth Environmental Protection Agency to prepare project-specific guidelines, and although public input is not required, it often occurs in practice (see Chapter 7). Scoping is available in about half the jurisdictions, but in others it is not a formal requirement. In some jurisdictions, such as South Australia, there is also an opportunity for public input into the guidelines document prepared by the independent Major Developments Panel.

The number and nature of the different levels of formal EIA vary from jurisdiction to jurisdiction (see Table 3.3). The most consistency occurs in relation to a formal EIS level, although this level of EIA has a different name in some states. Just over half the states have provision for a lower level PER document, and about the same proportion have a

Table 3.3 Levels of formal EIA in Australian states and territories

State or territory	*Levels of EIA*			
	Lowest level			*Highest level*
ACT	PA	PER	EIS	inquiry
NSW	SEE REF		EIS	inquiry
NT		PER	EIS	inquiry
Qld			EIS or IAS	
SA	DR	PER	EIS	
Tas	level 1	level 2 DPEMP	level 3 EIS, SECIS, or EMP	
Vic			EES (often includes inquiry)	
WA	CER	PER	ERMP	inquiry

higher inquiry level. While there is only one level of EIA in Queensland, there are four levels in the ACT.

Australian Capital Territory

In the Australian Capital Territory there are four levels of assessment: preliminary assessment (PA), a preliminary analysis of impacts that is used to determine the need for further assessment; PER; EIS; and inquiry (see Table 3.3). There is also provision for an intermediary round-table conference used to clarify and/or modify the proposal or to determine the need for further assessment. This round-table conference may be held at any stage of the process, including the government's assessment phase and should automatically be triggered prior to a PER or EIS (ACTPA 1992a). Guidelines outlining the issues to be considered in the EIA are set by the Environment Minister, in consultation with the relevant Minister, and are open to public inspection. There does not, however, appear to be any provision for public comment or revision of the guidelines.

New South Wales

In New South Wales there are three levels of assessment. The two lowest levels include the review of environmental factors (REF) and the statement of environmental effects (SEE) under Part IV (for proposals that are unlikely to significantly affect the environment). The more complex levels include the EIS and the Inquiry. In relation to the inquiry mechanism, New South Wales is relatively unique, with the establishment of independent, accountable, and statutorily appointed Commissioners of Inquiry (Bates 1995). The EIS, however, is the primary level of assessment, and is the responsibility of the proponent. In producing the EIS, the proponent is required to consult with the Director General of the Department of Urban Affairs and Planning (previously the Director of Environment and Planning), and to address both the Director's Guidelines and the Regulations (for Part V proposals) with regard to form and content. There is no provision for public comment at this early stage. There is, however, an increasing emphasis on 'planning focus meetings' and 'community focus meetings', both of which are used to identify key issues to be considered in the assessment (Stone 1997). There does not appear to be a formal public scoping stage for either the REF or the SEE.

Northern Territory

In the Northern Territory there are two key levels of assessment that may be required following consultation with advisory bodies and the responsible Minister. These levels are the PER (previously termed the Preliminary Environmental Report (PrER)) and the more detailed EIS. PrERs were previously used as an information basis on which to determine whether or not an EIS was required, although in practice this was often used as a de facto EIS. It was not, however, a public document, and the PER was introduced in response to an increasing demand for public accountability (Coopers & Lybrand 1994; Environment Institute 1990). Both the EIS and the PER are the responsibility of the proponent, although continued liaison with the department is actively encouraged (Coopers & Lybrand 1994; CCNT 1990). Draft guidelines are produced by the Minister and are finalised following a public review period of 14 days.

The Minister may also trigger an inquiry in the event that a matter cannot be resolved by normal EIA procedures, although this has not yet occurred in practice (DLPE 1995). Coopers & Lybrand (1994) also note that informal assessment may be undertaken by the department. In the event that a proposal is modified, the Minister must reassess the environmental significance of the action and assessment levels.

Queensland

Under the Queensland State Development and Public Works Organisation Act there is currently only one level of assessment: impact assessment study (IAS). The IAS is the responsibility of the proponent (often the responsible authority). However, Coopers & Lybrand (1994) note that because IASs require public exhibition some assessments are termed environmental studies to avoid public input. In terms of scoping, the Green Book (non-statutory guidelines outlining policy and administrative procedures for EIA) stipulates that where the development involves complex issues or conflicting advisory body advice, 'terms of reference' should be prepared by the responsible authority following consultation with relevant agencies. In this case, public involvement is discretionary and tends to vary depending on the activity assessed (Coopers & Lybrand 1994).

The EIS specified in the Local Government (Planning and Environment) Act, which is prepared by the proponent and submitted with the development application, is the only level of assessment required. However, an informal document, termed an environmental assessment report (EAR), has been created by Brisbane City Council for applications that do not trigger an EIS (Coopers & Lybrand 1994). Pursuant to section 8.2 (4–5) of the Local Government (Planning and Environment) Act 1990, the Chief Executive for Housing, Local Government and Planning is responsible (following government consultation) for setting terms of reference for designated developments. For non-designated developments, the local authority sets the terms of reference (section 8.2(12)). There is no formal public review mechanism for scoping, although a meeting may be convened for stakeholders (those directly involved or interested in the proposal) to identify relevant issues (Kinhill 1994).

South Australia

In South Australia the EIS was the only formal level of EIA until 1996 when an amendment to the Development Act introduced two new levels of assessment: the PER and the development report (DR). The 1996 amendment to the Act also created the Major Developments Panel, an independent body comprising six members appointed on the basis of different specified areas of expertise in industry, commerce, economic development, environmental conservation, and other areas relevant to the development being assessed (DHUD 1997). Once the Minister has made a declaration under section 46 of the Act, the development application or project proposal is then referred to the Major Developments Panel, which has the responsibility of determining the level of environmental assessment required (DR, PER, or EIS). The panel must prepare a public report describing the development or project and any significant issues, and must then invite submissions over a 20-business-day period. In order to determine the level of assessment, the panel must take into account criteria prescribed in the Development Act Regulations. These criteria are based on a set of draft national criteria developed through the Australia and New Zealand Conservation Council (ANZECC) and are discussed in Chapter 8.

The panel is also required to formulate guidelines for the preparation of the DR, PER, or EIS related to the proposed development or project. Scoping of the issues relevant to the assessment of the project is required, together with an indication of their priority. In addition, the panel has to consult the EPA(SA) on the preparation of guidelines if the development or project is a prescribed activity of environmental significance (as defined by the South Australian Environment Protection Act). Both the guidelines and the advice on the level of assessment are then given to the Minister, who is obliged to make them public and to advise the proponent.

Tasmania

The three Tasmanian assessment levels reflect the three mechanisms for triggering EIA described on pp. 45–7, and also tend to reflect documentation used prior to the new Environmental Management and Pollution Control Act. Level 1 assessments may only require the initial application form (for example, the development proposal), although in some cases further information may be necessary (see, for example, section 54 of the Land Use Planning and Approvals Act). Level 2 assessments may require the proponent to complete an environmental effects form or a development proposal and management plan. In the event that further information is required, formal guidelines may be set by the planning authority and the Board of Environmental Management and Pollution Control, pursuant to section 74 of the Environmental Management and Pollution Control Act (DELM 1996). Level 3 assessments may require formal documentation in the form of an EIS (prepared by the proponent) (DELM 1994) and/or a social, economic, and community impact statement (SECIS) (prepared on behalf of the Sustainable Development Advisory Council but paid for by the proponent) (DELM 1994). Alternatively the formal documentation required is an environmental management plan (EMP), sometimes called an impact assessment report (IAR).

While these documents are not explicitly referred to under the State Policies and Projects Act, they may be requested under section 20(3), which provides for 'matters to be addressed in the integrated assessment . . . the process to be followed . . . and the time within which the integrated assessment must be completed'.

Under section 74 of the Tasmanian Environmental Management and Pollution Control Act, it is up to the responsible authority to provide guidelines to the proponent regarding the potential environmental impacts, possible issues of public concern, the level of assessment, and the timing for each stage of the assessment. Before proceeding to level 3 assessments, councils or government agencies must be allowed to make comments on the project within 28 days of being notified that an assessment will take place. There does not appear to be any formal provisions for public involvement at this scoping stage, although in the past the practice was for level 2 guidelines to undergo an informal government consultation process for 4–6 weeks (DELM 1994). In this case, public involvement was discretionary.

Victoria

Only one level of assessment exists in Victoria: the environmental effects statement (EES). However, a public inquiry is often held at the completion of the public exhibition period (provided for under section 9 of the Environment Effects Act). Previously, preliminary environment reports (PrERs), similar to those in the Northern Territory, were also used to assist in determining the need for an EES, but these were omitted from the Act in 1994, being unnecessary in practice (DPD 1995).

Scoping of the issues and EES contents is the responsibility of the proponent, and is a public process with 28 days for public comment (although this public scoping process is not specified in the Act). A key feature in this process is the use of a consultative committee, established by the Minister for Planning, which guides both the scoping and the preparation of the EES. The committee is chaired by an officer from the Environment Assessment Branch of the Department of Planning and Development and may consist of the proponent, government agencies, local government, and community and environmental groups (DPD 1995). While not a legislative practice, the consultative committee process has been practised since mid 1990 (Environment Assessment Branch 1995).

Western Australia

In Western Australia once information is received from the proponent the EPA(WA) determines the required level of assessment under section 40 of the Environment Protection Act. A relatively unique feature of the Western Australian process is the system of 'multi-graded' assessment levels (Sammarco 1990). This system has a total of six assessment options, both informal and formal. These options are based on the 'location, size, complexity, range, duration, manageability and public interest of the proposal' (DEP (WA) 1993/4). Where effects are deemed to be insignificant by the Environmental Protection Authority (EPA(WA)), informal processes may involve no assessment, although guidelines may still be outlined (EPA (WA) 1993). Alternatively informal review may take place with non-binding public advice if the EPA(WA) deems impacts to be easily manageable (up to 65 per cent of all proposals fall into this category each year) (EPA(WA) 1993).

Formal documentation is not specified in the Environmental Protection Act. However, the Administrative Procedures established under the Act specify four levels of assessment:

1. consultative environment review (CER) for localised assessment projects with relatively manageable impacts
2. public environmental review (PER) for intermediate and regional scale proposals with potential for significant environmental effects or major public interest—PERs are also initiated following modifications to proposals that have already undergone the next highest level of assessment
3. environmental review and management program (ERMP), applied to statewide and strategic level projects—ERMPs incorporate the most detailed and comprehensive study (equivalent to the EIS), provide for more extensive public review, and provide a greater emphasis on environmental management to improve the development proposal (Snashall 1993; Bailey and English 1991; EPA(WA) 1993)
4. public inquiries—these may be initiated under section 40(2)(c) of the Act, but only with authority from the Minister for Environment, and only where proposals may have indirect consequences and numerous major impacts, and where public concern is significant at the state or national level (EPA(WA) 1993).

For each level of formal assessment, the EPA(WA) establishes guidelines based, in part, on government and public comment during a non-specified period (although public involvement is discretionary). In the event that a proposal is straightforward and is accompanied by sufficient documentation (this tends to happen with some CERs) the EPA(WA) does not have to issue guidelines (EPA(WA) 1993). A public record of assessment levels and other information required is established under section 39, although the EPA(WA) may, at the request of the proponent, exempt some proposals from this requirement.

CONCLUSION

As noted at the beginning of this chapter, screening is important to determine if projects should be subject to the EIA process. In Australia screening is usually a discretionary or subjective decision made either by a Minister with responsibility for the EIA portfolio or by a responsible authority. The use of prescribed lists of projects, actions, or criteria to trigger the EIA process is not common, with the notable exception of New South Wales.

It is at the scoping (or issue-focusing) stage that there is often the first opportunity for public involvement in the EIA process, although only about half of the various Australian jurisdictions provide for formal public involvement. Where it exists, the formal public scoping tends to avoid later criticisms of lack of early public involvement or inappropriate coverage of important issues. Various government departments also often conduct additional ad hoc scoping to focus on key decisions that have implications for government.

The formal levels and types of assessment appear quite complicated because of the plethora of different terms used for the various documents. In reality, there is an EIS-equivalent document in all states and territories (even though the name may be different). Many jurisdictions have a higher inquiry level of assessment, although this has not been used much in practice, and fewer jurisdictions have the lower level PER documentation. Within these broad parameters there is variation between systems and terminology, particularly for the lower levels of EIA documentation. However, given the importance of the EIS as a consistent level of formal EIA documentation, both in Australia and globally, this text uses the EIS as the main focus of its analysis.

4

EIS preparation and public participation

There is variation in the way an EIS document (or other documents relating to other EIA levels) is prepared, in the extent to which there is public participation during the process, and in the manner in which it is assessed. Some of the methods used in these three stages are discussed below, although it should be pointed out that the stage following the assessment of the EIS is the critical decision-making stage that EIA is designed to inform. Decision-making essentially comes after the formal EIA process has concluded, even though any decision to approve a project may set monitoring and auditing requirements based on the EIA report. For this reason, decision-making is discussed in Chapter 5, along with other EIA-related activities such as monitoring, auditing, and appeal rights, all of which follow the stage of formal EIA documentation.

EIS PREPARATION AND PRESENTATION

The content and quality of an EIS depend on a number of factors such as the guidelines or scoping of issues to be included, the extent to which there is liaison and consultation with the assessing authority, the willingness of the proponent to undertake appropriate studies, and the level of environmental expertise and professionalism of those preparing the report.

Although there is a standard content for an EIS, as discussed in Chapter 1, the content requirements vary between jurisdictions. Wood (1995) concludes his review of seven different EIA systems with the finding that only five out of the seven jurisdictions had satisfactory EIS content requirements and controls.

Responsibility for the preparation of the document varies from jurisdiction to jurisdiction. In California, for example, the EIS document may be prepared by the 'in-house' expertise of the staff of the agency taking the responsibility for the project; by consultants hired by the agency; by the proponent or the proponent's consultant; or by a consultant paid by the proponent, under the direction of the agency (Wood 1995). However, in Australia most EISs are prepared by consultants on behalf of the proponent or developer of a particular project. In some cases the proponent may also have the necessary expertise to prepare an EIS document—for example, if the proponent is a large firm of consulting engineers. The expertise of the consultant is a critical factor in the quality of the EIS, although few jurisdictions require accreditation of EIS consultants. In practice, there is generally a pool of consultants who tender for EIS contracts. The competition generally has the effect of eliminating poor performers, unless there are other factors such as politically favoured companies or linkages between design/planning stages and construction contracts.

Glasson, Therivel, and Chadwick (1994) discuss the contents of an EIS, its organisation, its clarity of communication, and its presentation. They discuss best practice in terms of these factors and conclude that an EIS should:

- be comprehensive
- explain why some impacts are not addressed
- emphasise key points
- set the context of the issues
- have a non-technical summary
- be a unified document
- be as brief as possible while presenting necessary information
- be well written and include a minimum of technical jargon
- clearly state any assumptions
- be specific about impacts and quantify these where possible
- be honest and unbiased.

They also discuss aspects of the presentation of the final document and comment on the variation in document length and quality. Jones et al. (1991) note that in the United States EISs are expected to be less than 150 pages in length but in the United Kingdom 15 per cent of EISs exceed 150 pages.

A major issue for public presentation of the EIS document is its quality and content, both of which should use the best-practice principles outlined by Glasson, Therivel, and Chadwick in order to provide the public with an adequate document for comment. For example, the lack of specificity in describing either the project or its impacts may result in an EIS that is more conceptual than descriptive and that lacks the substance for realistic comment or debate. It is important that there is a mechanism for checking the quality of the EIS prior to its release. Wood (1995) is critical of both the UK and New Zealand systems for lacking such formal checking mechanisms. In Australia the checking is usually done by an environment assessment branch of the relevant government agency prior to the public release of the EIS.

EIS preparation and presentation in Australian states and territories

In Australia, EIS length and quality varies between states and territories, but the greatest variation occurs according to the project complexity and the style of presentation of the EIS. This quality and style is usually determined by the proponent's willingness and financial support in conducting appropriate environmental investigations, together with the professionalism of the EIS consultant. Very little research has been conducted in Australia into the quality of EIS preparation and presentation, although there are some anecdotal observations.

Gilpin (1995) refers to shortcomings in EISs prepared in New South Wales, which were identified by the New South Wales Department of Environment and Planning (1995), the New South Wales Science and Technology Council (1983), and Smyth (1987). A number of these shortcomings refer to common problems in EIS preparation, such as:

- too much technical information
- poor baseline scientific data (information on existing environmental conditions)
- lack of appropriate scoping
- a narrow focus on engineering data rather than broad environmental data
- failure to justify the project and discuss alternatives
- failure to address cumulative and indirect impacts.

There were also criticisms of the poor presentation of material, unclear text, use of technical jargon, and incomprehensible diagrams and figures. Gilpin (1995) suggests that from the mid 1970s to the mid

1980s about 25 per cent of EISs in New South Wales were inadequate for the purposes of public exhibition and would have been unable to withstand scrutiny through public inquiry. Smyth (1987) is more critical and suggests that over the same period 75 per cent of the EISs were inadequate for the purposes of public communication and effective participation, and that over 50 per cent of the EISs fail to adequately address environmental issues relevant to decision-making.

In South Australia, as noted by Harvey (1995), there are major differences in the presentation of EISs. Presentation ranges from short, typed, photocopied, and stapled documents, such as the Remove All Rubbish EIS (1983), to professionally produced, glossy, coloured, spiral-bound and printed documents, such the Jubilee Point EIS (Jubilee Point Pty Ltd 1986a). However, the presentation is not necessarily a reflection of content: the Waterloo Corner EIS was considered an adequate document, whereas the Jubilee Point EIS received extensive public criticism. Harvey (1995) notes examples of sub-standard EISs, including one produced by the South Australian government for a project in which it was a joint partner.

The length and public accessibility of some EISs has also become an issue. In contrast to the US expectation that EISs should be less than 150 pages in length, a number of Australian EISs are voluminous and expensive. The EIS for the Multifunctionpolis(MFP) Australia in South Australia (OPUD 1992) is a glossy, spiral-bound document with numerous coloured maps and illustrations, and it runs to over 400 pages, including appendices. Similarly, the Sydney Third Runway EIS runs to 720 pages (FAC 1990). Another problem with these more expensive EIS documents is the need for a subsidy on the purchase price to make it more accessible to the community. In South Australia, for example, the practice has been to keep the document price within the $20–30 price range for even the larger documents such as the MFP Australia EIS.

Consultation and Public Participation

As noted by Wood (1995, p. 5) consultation and public participation should be important inputs at each stage in the EIA process, although the sorts of people and bodies invited to comment may vary from proposal to proposal. There has been debate on the need to improve public participation in EIA (Clark 1995) and on the theoretical aspects of public participation in environmental decision-making (Robinson 1994). In the United States, public debate and litigation have been important in the evolution of EIA procedures, whereas the relatively recent (1988) introduction of EIA procedures into the United Kingdom was set against a traditional background of British administrative discretion and secrecy, with limited public input (Glasson, Therivel, and Chadwick 1994).

Consultation often occurs with government departments and various agencies at a number of stages through the EIA process—stages such as screening, scoping, and report preparation—but public consultation at the early stages is less common. However, once the EIS has been prepared, most jurisdictions have a formal public participation process that can be implemented using a number of methods, such as:

- display period for the EIS and opportunity for public response
- public information meetings and displays
- on-site information centres or telephone hot-lines
- questionnaires and surveys
- advertisements in the media.

Whatever methods are used, the aim of public participation in the EIS process is to make sure that the public can comment on environmental issues in the EIS, or can raise new issues, which can then be considered in decision-making. In some jurisdictions, particularly the United States, public participation is also associated with appeal rights over EIA-related decisions. In his survey of seven different EIA systems, Wood concludes that consultation and participation prior to and following EIA occur fully in only three out of the seven EIA jurisdictions (Wood 1995).

Public participation and consultation in Australian states and territories

In Australia, the role of public involvement is well accepted. The National Approach to EIA (ANZECC 1991b) outlines agreed principles for public involvement in the EIA process. Similarly, an intergovernmental agreement on the environment, *Intergovernmental Agreement on the Environment* (IGAE 1992), contains a provision for public involvement in the EIA process (see Chapter 8). The opportunities for, and timing of, public participation in the EIA process vary between states and territories, but as noted by Harvey (1994a, 1995), it is important to examine the role of formal and informal public involvement in the EIA process by examining EIA in practice. Formal public involvement in the EIA process usually takes place through public submissions on EIA documents, although this is not necessarily the most effective form of public input.

Table 4.1 Public review of EIA documents in Australian states and territories

State or territory	*Length of time for public review of different levels of EIA documents*			
ACT	PA—for approved period	PER discretionary	EIS—28 days	inquiry
NSW	SEE—14 days REF—14 days	maximum	EIS—minimum 30 days	inquiry
NT		PER maximum 28 days	EIS—minimum 30 days	inquiry
Qld			IAS—discretionary (but 60 days in practice)	
			EIS—30 days (for applications requiring public notice); discretionary for others	
SA	DR—minimum 15 days	PER—minimum 30 days	EIS—minimum 30 days	
Tas.	level 1—14 days	level 2	level 3[1]	
		DPEMP—discretionary, but 14–28 days in practice	EIS, SECIS, IAR—not specified, but 28 days in practice	
Vic.			EES—not specified in Act, but 2 months in practice	inquiry
WA	CER—4 weeks	PER—8 weeks	ERMP—10 weeks	inquiry

1 Note that the public may also review the Sustainable Development Advisory Council's assessment report for 28 days or longer.

Public involvement in Australian EIA occurs at different stages of the process. These stages include:

- the triggering of an EIA
- public scoping of environmental issues and guidelines
- public involvement on consultative committees
- public inquiries
- appeal (although involvement at this stage is limited).

However, there is no consistency between Australian states and territories in the provision of formal public involvement in the EIA process.

Australian Capital Territory

In the Australian Capital Territory the PA is placed on public exhibition for an approved period, during which time the proponent may hold public meetings (Norman 1997). There is no time period for public review of the PER outlined in the Land (Planning and Environment) Act. However, the Environment Minister may require the proponent to undertake consultation during preparation of the report, and to make the finalised report available for public inspection (section 124). There does not appear to be any provision for written submissions or proponent response. In contrast, the EIS is available for a minimum of 28 days and may be revised by the proponent in light of written comments. The public are also able to access reports from round-table meetings and from an inquiry report (if an inquiry has been held). However, section 133 states that any information that is not in the public's interest or that breaches proponent confidentiality may be excluded from public review.

New South Wales

Under the New South Wales Environmental Planning and Assessment Act, following completion of the EIS document, the consent authority (Part IV) or the determining authority (Part V) releases the EIS (plus the development application for Part IV proposals) for public review for a minimum period of 30 days (Department of Planning 1985). At the end of the period, and not less than 21 days before the decision is made, all submissions are forwarded to the Department of Urban Affairs and Planning for consideration (DEP (NSW) 1985). Like the Northern Territory process, any information considered confidential may be withheld from the public by the determining authority (Part V proposals only) (Farrier 1993; Environment Institute 1990). In addition, the proponent, in formulating a final EIS, is not required to respond in any way to the public submissions, although this is often undertaken anyway (Bates 1992; Environment

Institute 1990). In the case of REFs, the document may be exhibited for 14 days while the SEE is exhibited, with the development application, for the same period.

Northern Territory

In the Northern Territory once the PER is completed it is placed on public exhibition for a maximum of 28 days; the EIS is reviewed for a minimum of 28 days. However, there are provisions in the Environment Assessment Act that may restrict public review of the PER or EIS under paragraph 10 of the Administrative Procedures. These provisions allow proponents, in certain circumstances, to object to the release of information for public review. Proponents lodge an objection to this effect with the Minister for Environment, and the Minister must consult with the responsible Minister before making a decision not to release information. At the end of the public review period, the proponent who prepared a draft EIS must produce a supplement to the document. This supplement, described as the final EIS, considers and outlines responses to written public comments and inquiry results (if an inquiry has been conducted). The process of producing a supplement report is not required for the PER level of assessment.

Queensland

In Queensland there are no specific requirements for public review of the IAS in the State Development and Public Works Organisation Act. However, following consultation with the proponent and advisory bodies, the responsible authority may release the IAS for public review for an unspecified period—although Kinhill Engineers (1994) note that a review period of 60 days usually occurs in practice. Public review is also encouraged in the Green Book (Co-ordinator General, Premier's Department 1987). Advisory bodies listed in Schedules 1 and 3 of the Act may include non-government organisations, academic institutions, or special interest groups (special interest groups must be listed under Schedule 1 of the Green Book—for example, the Main Roads Department), but overall, public review is discretionary and there does not appear to be any explicit requirement for the proponent to respond to comments or to amend the IAS in response to public comment. This does, however, occur in practice: a supplement is prepared, which, once completed, is distributed to all advisory bodies that made substantive comments (Queensland Department of Environment, Planning and Assessment Branch, pers. comm., May 1997).

Under section 8.2 of the Queensland Local Government (Planning and Environment) Act, public review is only provided for designated developments that require the giving of public notice (section 8.2(7))—for example, rezoning or consent applications (Scanlen 1993). In this case, a period of not less than 30 working days is set for public objections. During this time the EIS is also referred to government departments for comment Queensland Department of Environment, Planning and Assessment Branch, pers. comm. May 1997).

In other cases, public review is discretionary (Scanlen 1993) and there does not appear to be any requirement for the proponent to respond to objections, as occurs in practice under the State Development and Public Works Organisation Act.

South Australia

In South Australia there is opportunity for the public to comment to the Minister on the EIS, the PER (this comment period must last a minimum of 30 business days, beginning from when the document is publicly released) or the DR (this comment must last a minimum of 15 business days, beginning from when the document is publicly released). There is a requirement for a public meeting to be held during the public exhibition of the EIS and the PER but not the DR, although there does not appear to be a specification for recording comments at these meetings and taking account of issues raised (apart from the official public written submissions to the Minister).

Following the public review period, the proponent must respond to comments made on the EIS and the PER, but is not compelled to respond to comments on the DR.

Tasmania

Each level of assessment in Tasmania is subject to public review under section 74(6) of the Environmental Management and Pollution Control Act, although to varying degrees, given that time frames are only specified for level 1 activities. In addition, section 60 of the Land Use Planning and Approvals Act requires both level 1 and 2 activities to refer permit applications for the 'use or development of land' to relevant agencies for a comment period of 28 days or longer.

In terms of public comment, level 1 undergoes the normal objection period of 14 days or less that is required for licence applications under the Land Use Planning and Approvals Act (section 57(5)), although this is discretionary (section 57(2)). Previously, level 1 assessments underwent a review period of 30 days (DELM 1994). There does not appear to be any formal requirement for the proponent to respond to submissions. The public review time period for level 2 activities appears to be discretionary given that there is no specified period within the Environmental Management and Pollution Control Act, and that time frames under the Land Use Planning and Approvals Act do not apply at this level. Again, the proponent is not required to respond to submissions.

As is the case with level 2, there is no specified period for public review of the EIS (or equivalent) for level 3 activities, under either the Environmental Management and Pollution Control Act or the State Policies and Projects Act, although in practice the public review period has been 28 days. There is, however, a unique feature at this level: the public may comment on the government's draft assessment report for 28 days (or longer as specified by the Minister) (section 22 of the State Policies and Projects Act), although this provision may be omitted by the Minister under section 20(4).

Victoria

In Victoria the Guidelines require that, prior to its public release, the EES is checked by the Department of Infrastructure for suitability in terms of content and form. Although provided for under section 9(2) of the Environment Effects Act, the public review period for the EES is not specified in the Act. The Guidelines, however, state that the EESs are usually displayed for two months (Department of Planning and Development 1995). Following this public review period, the proponent is required by the Guidelines to respond to submissions (public and government) at an inquiry panel, when held, or to the Chief Assessment Officer. The response is usually outlined in an appendix to the Minister's assessment report if no inquiry is held (Department of Planning and Development 1995). At this stage, the Minister for Planning may also call for a supplementary statement (section 5), which is subject to the same public exhibition requirements as the EES (DPD 1995). However, this provision is rarely used in practice (DPD, pers. comm., 1995).

Western Australia

Under the Western Australian legislation, draft documents prepared by the proponent are examined by the EPA(WA) in terms of their suitability for public review. If the EPA(WA) deems them inadequate, it may require additional material or a revised draft of the assessment (EPA(WA) 1993).

Once approved, the document is placed on public exhibition (with the exception of confidential information) for a period of four weeks (CER), eight weeks (PER), or ten weeks (ERMP). The EPA(WA) then summarises submissions (including government comments), and the proponent subsequently responds to comments and may amend the proposal. This response by the proponent apparently occurs at *all* assessment levels, which is significant. In other jurisdictions, the lower levels of assessment (for example, PER) generally do not require a response.

CONCLUSION

In contrast to some overseas research, there appears to be a lack of detailed research into the content and quality of Australian EIS documents. In addition some countries such as the United Kingdom have a professional EIA Institute geared towards quality control in documentation and in providing assistance in the professional evaluation of completed documents. Although some Australian state and territory jurisdictions hold databases of EIA consultants, the quality of EIS documents is largely left to the professional competence of the consultants, many of whom specialise in EIS work. Quality control of EIS documents also varies between jurisdictions, and consequently, there is considerable variation in the quality and content of EISs.

The extent of formal public consultation also varies between jurisdictions, and in some cases the public exhibition and formal requirements for proponent response are discretionary: this is not the case in Western Australia, where the proponent has to respond to public comments for all EIA levels. In all jurisdictions most documents equivalent to an EIS are publicly exhibited for at least a month. What is not discussed in this chapter, but is picked up in the discussion of EIA practice in Chapter 6, is the fact that there is often considerable ad hoc public involvement and consultation occurring outside the formal provisions for public comment—for example, the EIS exhibition period. In addition, consultation with government departments and various agencies usually occurs at a number of stages throughout the EIA process. This tends to facilitate government involvement in the later assessment and decision-making stages.

5

Assessment, decision-making, monitoring, auditing, and appeal rights

Once an EIS has been completed and made public, the document is subject to a formal review, which varies depending on the procedures in the respective jurisdictions. In some countries such as Canada and the Netherlands, independent EIA commissions review the EIS, whereas in Australia the review is conducted by government agencies. As indicated in the previous chapter, there is usually a duty on the part of the proponent to respond to comments raised on the draft EIS, although the UK system has been criticised for lacking this requirement. In the United States and Australia it is the draft EIS that is reviewed and then a final EIS is prepared after review. In practice, most Australian final EISs are not a completely revised document but comprise the draft EIS together with agency comments, public submissions, and the proponent's response—these last three are published as a separate document called a 'supplement'. In Australia, the review of the final EIS is usually referred to as an 'assessment report' and is published as a separate document. Since it is largely government and government agencies involved in review and decision-making, this chapter focuses on the government input to the EIA process, unlike the previous chapter that focuses on the input of the proponent, the consultant, and the public.

ASSESSMENT (OR REVIEW) AND DECISION-MAKING

According to Sadler (1996), formal EIS reviews focus on common aspects, including what he terms the 'triple-A test' of 'appropriateness, adequacy and actionability'. Appropriateness relates to the coverage of key issues and impacts, and conformity with study objectives. Adequacy relates to the reliability of data analysis, interpretation, and verifiable scientific methodology. Actionability refers to the relevance of the EIS for decision-making, including a clear description of consequences and relevant management options. There have been attempts to develop criteria for EIS review by authors such as Lee and Colley (1990), who propose a hierarchical review method based on gradings (from A to F) allocated to four main components of the EIS. These components are:

- description of the development, local environment, and baseline conditions
- identification and evaluation of key impacts
- alternatives and mitigation of impacts
- communication of results.

Each of these is in turn split up into more specific categories. Similarly, Tomlinson (1989) proposes review criteria for nine main EIS components, although his method uses yes/no answers rather than grades. Glasson, Therivel, and Chadwick (1994) provide their own comprehensive checklist of EIS review criteria based on some of the earlier studies.

A problem in the review or assessment of EISs is maintaining objectivity. One way of attempting to achieve this is by the use of the above review criteria method, a popular method in the United Kingdom. Another method is to use independent review consultants. Perhaps the most 'objective' method is the use of review panels that are independent from the decision makers. This appears to be best achieved in Canada and the Netherlands, where the review panels are seen as authoritative, publicly accountable, and influential.

An important factor in the compilation of the assessment report is the use of an interdisciplinary team of experts and the use of holistic methods of assessment that consider environmental linkages and cumulative or transboundary impacts. Australian assessment reports are mostly produced by such teams within government environmental impact assessment branches. However, these assessment reports are not necessarily the key determinant in the decision-making that follows the EIA process. Similarly, Wood (1995) notes that six out of the seven EIA jurisdictions he reviewed did not require either the EIA report or the review to be a central determinant of decision-making.

Assessment and decision-making in Australian states and territories

In Australia most reviews or assessments are conducted using in-house government expertise, and even the Commonwealth Environmental Protection Agency, which was part of a broader environment

Table 5.1 EIA assessment and decision-making in Australian states and territories

State or territory	*Assessment responsibility*	*Decision-making*
ACT	Environment Planning and Assessment section of the Australian Capital Territory Planning Authority	Legislative Assembly
NSW	Department of Urban Affairs and Planning	Pt IV—consent authority
		Pt V—determining authority (but may only proceed with consent from the Minister for Urban Affairs and Planning)
NT	Environment Assessment Branch of the Environment Protection Division of the Department of Lands, Planning and Environment	responsible Minister
Qld	State Development and Public Works Organisation Act—Co-ordinator-General (Department of Environment) and responsible Authority	State Development and Public Works Organisation Act—responsible authority
	Local Government (Planning and Environment) Act—Department of Housing, Local Government and Planning, and local government authority	Local Government (Planning and Environment) Act—local government authority
SA	EIA Branch of the Department of Housing and Urban Development	Governor
Tas	level 1—local council/planning authority	level 1—local council/planning authority
	level 2—Board of Environmental Management and Pollution Control	level 2—Board of Environmental Management and Pollution control
	level 3—Sustainable Development Advisory Council	level 3—Governor
Vic	Office of Planning in the Department of Infrastructure	relevant Minister
WA	EPA(WA) and the Department of Environment Protection	Minister for Environment (using 'whole of government' approach)

department, appears to have been less of an 'independent arbiter' than Wood (1995) suggests.

Australian Capital Territory

Pursuant to section 127 of the Australian Capital Territory Land (Planning and Environment) Act, the final EIA document is forwarded to the Environment Minister together with a report on the consultation process and a summary of each public submission received. In terms of the PA if no further assessment is required, the process ends with the incorporation of the Minister's (either the relevant Minister or the Environment Minister) evaluation into the normal application process (ACTPA 1992).

Assessment of the PER and EIS is more time consuming, often requiring a preliminary review prior to full evaluation (within 42 days of submission to the Minister). At this stage, under section 128 of the Australian Capital Territory Land (Planning and Environment) Act, a meeting to clarify the issues (and any reports from an inquiry, if held) can be held with the proponent and any other person. A report of this meeting is publicly available and includes recommendations of the Environment Minister, which will be included in the final Evaluation Report (see below). The proponent may be required to provide further information or to revise the document for re-submission to the Environment Minister. Following (re)submission, the Environment Minister is required to produce a publicly available evaluation report within 56 days of receiving the document. The evaluation report outlines the appropriateness of the PER or EIS, comments on any impacts, and makes recommendations. The Minister may either recommend the refusal of the proposal (with justification) or offer approval with or without conditions. These conditions can include mitigation measures (ACTPA 1992). This report (or the inquiry report), together with the PER or EIS documents, is then tabled at the Legislative Assembly for final decision within six days.

New South Wales

Following public review of the EIS documents the Department of Urban Affairs and Planning considers all submissions (Part IV of the Environmental Planning and Assessment Act) in order either to determine whether the consent authority can proceed with the decision or to identify:

- the need for Minister interest (particularly under section 101 in relation to applications with state or regional significance)

- the need for a public inquiry (in which case, the consenting authority is replaced by the Minister as the decision maker)
- any matters that should be regarded by the consenting authority (Department of Planning (NSW) 1985).

Although there is no provision for an assessment report such as that required for Part V of the Act, these are prepared, in practice, by the department (Farrier 1993). The consenting authority (usually a local council) then makes a final decision based on criteria outlined in section 90. This determination includes the EIS, any submissions, and the proponent's response (Department of Planning 1985; Coopers & Lybrand 1994). However, in matters of state or regional significance—such as major employment-generating industries—or where an inquiry has been held under section 119, the final decision is made by the Minister. The lower-level SEE undergoes a similar assessment process, although it is the consent authority rather than the Department of Urban Affairs and Planning that considers public submissions before making a final decision (Department of Urban Affairs and Planning, pers. comm., 1997)

In regard to Part V of the Act, the Department of Urban Affairs and Planning may either undertake to produce an independent and public 'director's examination' (section 113(5)) for consideration by the determining authority (usually the proponent) or call for an inquiry (section 114). In the latter case, the Minister may give advice, and the Director General of the Department of Urban Affairs and Planning may also prepare an independent report if necessary. The determining authority is the final decision maker and prepares a final report (regulation 64) that is made available to the public. However, if the determining authority is also the proponent, the proposal cannot be approved without consent from the Minister for Urban Affairs and Planning, a new requirement under the 1993 amendments to the Act. In this case the determining authority requests approval from the Minister; an independent Director-General's assessment report may be produced, and the Minister for Urban Affairs and Planning makes the decision. The determining authority may then decide to proceed or not, in accordance with this decision (Department of Urban Affairs and Planning, pers. comm., 1997). This more complicated process does not occur for the lower level REFs. In the case of REFs, public submissions are considered by the determining authority before it makes a final decision. No independent assessment report is produced (Department of Urban Affairs and Planning, pers. comm., 1997)

Northern Territory

In the Northern Territory, the PER or the final EIS is examined by the Minister within 28 or 35 days (respectively) of public review, during which time the Minister consults with relevant advisory bodies, and may either request further information from the proponent (in the first 21 days of the 35-day assessment period for the EIS), or may request advice, from other bodies on the proposed action. Subsequently, the Minister makes non-binding recommendations based on written comments, other advice, and the results of an inquiry (if held), within a period of 14 days after the examination period (PER), or within the examination period or longer as specified by the Minister (EIS). Criteria for evaluating reports have also been produced by the Conservation Commission of the Northern Territory (1992). In practice, these Ministerial recommendations are based on an assessment report prepared by the Environment Protection Division of the Department of Lands Planning and Environment (DLPE 1995). The final determination on the proposal is made by the responsible Minister.

Queensland

On completion of an IAS under the State Development and Public Works Organisation Act, and after public review has been finalised (where it has been undertaken), the IAS is forwarded to advisory bodies for comment. No time period is specified, but comments are required to be made 'as soon as possible' (Co-ordinator-General, Premier's Department 1987). Although Bates (1995) notes that there is no final assessment process (for example, by means of an assessment report), a review report is prepared, in practice, by the responsible authority based on the IAS, the supplement (if provided), and on any advisory-body comments. Recommendations outlined in this report are then referred to decision-making authorities (Assessment Section, Department of Environment and Planning, pers. comm., May 1997). The final decision is made by the responsible authority after 'appropriate and responsible' consideration of the impacts of the development (Co-ordinator-General, Premier's Department 1987).

In the case of designated developments under the Queensland Local Government (Planning and Environment) Act, the completed EIS must be forwarded for comment (within a minimum of 20 working days) to the Director of the Department of Housing, Local Government and Planning, and to any referral agency nominated in the terms of reference (section 8.2(8)). Any comments, received from the Director on amendments to planning schemes, on staged rezoning, or on town planning consents are taken to be objections (Weir 1994; Coopers & Lybrand 1994; Haydon 1991). Subsequently, an internal planning report is prepared by the local government authority (although this requirement is not specified in the Act), which then makes the final decision within 60 days. In doing so, local government must only have 'regard' to the EIS (section 8.2(11)). However, if the proposal is of state or regional significance, the state government may play a role in the assessment by default (Department of Housing, Local Government and Planning, pers. comm., 1995).

South Australia

Under the South Australian Development Act (section 46(9)), 'the Minister [for Housing and Urban Development] must prepare a report (an "Assessment Report") that sets out or includes' an assessment of the development or project and comments on the EIS, on public submissions, on the proponent's response, and on any comments from the EPA(SA) or other relevant bodies. In practice, assessment procedures in South Australia are conducted through the EIA branch of the Department of Housing and Urban Development. The EIA branch coordinates comments from other government departments and agencies and then provides the Minister with a draft assessment report for authorisation.

All decisions on development applications subject to an EIS, PER, or DR under the Development Act are made by the Governor of South Australia (section 48), unless the Governor decides to delegate decision-making authority to the independent Development Assessment Commission (section 48(8)). In practice, the Governor takes advice from the Minister or Cabinet. The Governor's authority applies to both public and private developments, and the Governor must not grant approval until the environmental assessment has been completed. In making a decision, the Governor must have regard to:

- the EIS, PER, or DR
- the provisions of the development plan
- building rules (if relevant)
- the planning strategy
- matters relating to the Environment Protection Act.

However, there is also a provision in the Development Act (section 48(2)) for the Governor to refuse the application at any time during the EIA process. Decisions on projects that are not defined as 'developments' are made by the relevant authority.

Tasmania

In Tasmania there does not appear to be any formal assessment report for level 1 or level 2 activities. In the case of level 1, the final decision is made by the relevant local government or planning authority within 42 days of receipt of the application under section 57(6) of the Land Use Planning and Approvals Act. However, if the application (requesting development consent) has been referred to government agencies for comment, this period is extended to 60 days. In terms of level 2 activities, the decision and any conditions are made by the Board of Environmental Management and Pollution Control under section 25(5) of the Environmental Management and Pollution Control Act. Any refusal must be complied with by the planning authority. However, if the board permits the activity, the planning authority may still refuse the application (Townsend, pers. comm. 1996). Conditions set by the board may include a requirement for an environmental management plan, and monitoring and rehabilitation requirements (section 25(6)).

Level 3 is substantially different and is a relatively unique process in Australia. Once the EIS (or equivalent) is complete, the Sustainable Development Advisory Council established under the State Policies and Projects Act prepares a draft integrated-assessment report in consultation with government agencies. As noted previously, this is publicly reviewed for 28 days, and the Sustainable Development Advisory Council is required to respond to public submissions. In this case, they may hold a hearing under section 24, and modify the report to form the final integrated-assessment report that is publicly available, which makes recommendations to the Minister regarding conditions and whether or not the proposal should proceed. The Minister or the Governor (on Ministerial recommendation) must then make a determining order within 28 days, which may either accept the report as a basis for decision or substitute an alternative decision (in which case it must be approved by both houses of parliament) (DELM 1994).

Victoria

Assessment of the Victorian EES, of any submissions, and of any results from inquiries (if they have been held) is made, in an assessment report, by the Office of Planning in the Department of Infrastructure, on behalf of the Minister for Planning. This assessment report is publicly available (although, again, this has no legislative status). While no time period for assessment is specified in the Act, the Guidelines state that assessment should be completed within three months of the public exhibition period, or one month after an inquiry report. As is the case for the majority of states and territories, the assessment is only advisory and the Office of Planning cannot make any legally binding conditions or decision. However, in 1995 the Environment Assessment Branch (now the Office of Planning) stated that advice has rarely been rejected in practice (Environment Assessment Branch, Department of Planning and Development 1995).

After considering the assessment report, the final decision is made by the relevant Minister. This decision may take one of five key decision-making avenues: planning scheme amendments, planning permits, mining licences, works approvals, or other government decisions. In the event that the recommendations of the Environment Assessment Branch are not adopted in the final decision, the Guidelines require the Minister who has made the decision to prepare a written and public statement of justification.

Western Australia

In Western Australia, the proposal is assessed by the EPA(WA), which makes an assessment report publicly available within six weeks of the assessment. The report incorporates recommendations to the Minister for Environment. These recommendations are based on environmental factors, the public issues raised, an inquiry (if held), government agency advice, the proponent's response, and EPA(WA) research (EPA(WA) 1993–94, 1993). It is of interest that the assessment by the EPA(WA) may proceed without the proponent's response to public submissions if a 'reasonable' time has passed (EPA(WA) 1993). The final advice to the Minister included in the report may include whether or not the proposal is 'environmentally acceptable' and any conditions or procedures to which implementation of the proposal should be subject (EPA(WA) 1993–94). Importantly, the Minister has the power to direct the EPA(WA) to assess, or reassess, the proposal more fully or more publicly (section 43 of the Environmental Protection Act).

A joint decision-making process is then initiated under section 45 of the Act. This process involves a 'whole of government approach', whereby the Minister for Environment consults with relevant Ministers and other decision-making authorities to determine whether or not the project should proceed and, if so, what conditions should be set (Bailey and English 1991, 1994). The final consultation and decision usually takes four weeks, and all conditions set by the Minister for Environment are legally binding. The legally binding nature of these conditions is a unique feature that is absent in other states and territories, where recommendations usually only have advisory status. In the event there is a disagreement between Ministers, the Governor may make the final decision or establish an appeals committee under section 45 of the Act.

MONITORING AND AUDITING

Although not strictly defined in this book as part of the formal EIA process, the implementation of the monitoring and auditing programs that follow the approval of an EIS-related proposal are usually developed from or based on the outcomes and recommendations of the EIS and its assessment. It is here that Wood's (1995) review found a major weakness, with none of his seven jurisdictions taking the monitoring of action-impacts fully into account or linking it to earlier stages of the EIA process. It is clear that maximising the effectiveness of monitoring entails developing greater links between the decision makers and the EIS process. Glasson, Therivel, and Chadwick (1994, p. 166) suggest that 'EIA should not stop at the decision' and should build monitoring and auditing into the process. Similarly, Sadler (1996) suggests that unless there is a minimum follow-up capability for EIA, there is a danger that, the process becomes a pro-forma exercise rather than a meaningful mechanism for environmental management.

There has been some confusion over the use of the term 'monitoring' in environmental science (Harvey 1981), but the use of monitoring and auditing in EIA is becoming better defined (Bisset and Tomlinson 1988; Glasson, Therivel, and Chadwick 1994; Sadler 1996; Wood 1995). Monitoring in EIA is generally referred to as either implementation or impact monitoring, both of which employ a variety of methods depending on the environmental parameters to be measured.

Implementation monitoring checks if the project or action has been implemented in accordance with the conditions of the approval, with particular reference to environmental impact mitigation measures. This type of monitoring is usually the responsibility of planning or building authorities under the relevant legislation. If the approval conditions are not met, alternative mitigation measures should be introduced and in some cases the project may be stopped altogether.

Impact monitoring is usually more specifically related to environmental impacts resulting from the implementation of the project or action—for example, the impact of breakwater construction on the long-term littoral drift of sediment; or the introduction of weeds resulting from the clearance of native vegetation. Impact monitoring is usually conducted by environmental authorities, who are sometimes assisted, in financial or real terms, by the developer. However, impact monitoring is focused on collecting and analysing environmental data, and is, therefore, not necessarily linked to further action or environmental mitigation.

In the context of EIA, the term 'auditing' can be defined in different ways. Tomlinson and Atkinson (1987) refer to seven points of audit in the EIA process, whereas Buckley (1991) refers to two main types of auditing: environmental impact auditing and environmental management auditing. Buckley suggests that 'impact audits provide a means for both industry and government to demonstrate their competence in environmental management to the public' (Buckley 1991, p. 21). Impact auditing compares the predicted impacts and mitigation measures from the EIA with the actual impacts and mitigation measures that occur once the project or action has been implemented. Environmental management auditing, on the other hand, refers to the performance of corporate programs and structures for environmental management and assessment of corporate risks and liabilities. Wood (1995) provides a summary of the main conclusions from the literature on impact auditing, which comprises the greatest part of the literature on EIA monitoring and auditing.

Main conclusions about impact auditing in EIA

- There are no standardised audit methodologies.
- Monitoring needs to be considered early in the EIA process.
- Post-EIA report design changes invalidate many forecasts.
- Many EIA reports contain few forecasts.
- Many EIA report forecasts are vague and qualitative.
- Monitoring is often inadequate for auditing purposes.
- Only a minority of EIA forecasts have been accurate.
- Few EIA report forecasts have been totally inaccurate.
- Few unanticipated impacts have been detected.
- There is little evidence of systematic bias in forecasts.

Wood 1995, p. 200

MONITORING IN AUSTRALIAN STATES AND TERRITORIES

In Australia, little attention has been paid to monitoring and auditing methods, and although there are provisions for monitoring in a few jurisdictions, these provisions are largely discretionary and have not been an important element of the EIA process. Buckley (1989) conducted what he referred to as the first audit of environmental impact prediction in Australia, and he is critical of the lack of verifiable predictions in the EISs he used.

Australian Capital Territory and New South Wales

In the Australian Capital Territory, and in New South Wales, the legislation does not appear to contain any explicit EIA requirements for monitoring of the proposal, although the Australian Capital Territory Land (Planning and Environment) Act does provide for a public register containing information on all proposals subject to EIA. Moreover, in practice, monitoring is required in New South Wales for all approvals made by the Minister. Guidelines also recommend that monitoring conditions are outlined in the environmental management plans prepared for assessed projects, and that the Department of Urban Affairs and Planning conduct performance audits of all Ministerial approvals (Department of Urban Affairs and Planning, pers. comm., 1997).

Northern Territory

In the Northern Territory, paragraph 15 of the Administrative Procedures contains provisions for review and monitoring of safeguards, and of environmental effects. These provisions are essentially used as an audit tool to determine the effectiveness of the EIA process and to make further recommendations for environmental protection to the responsible Minister. These provisions are complemented by the newly established public register documenting the outcomes of project assessments.

Queensland

In Queensland there does not appear to be any formal EIA provisions for monitoring of the proposal following assessment (unless specified in any conditions of approval arising from the assessment) in either the State Development and Public Works Organisation Act or the Local Government (Planning and Environment) Act. However, in practice there is increasing monitoring occurring (Department of Environment, pers. comm., 1997).

South Australia

Monitoring was introduced in South Australia for the first time in 1996 through an amendment to the Development Act. Section 48C of the Act makes provision for the Minister to require a person undertaking or benefiting from a development or project that has been subject to the EIA requirements to conduct monitoring and/or to comply with an audit program. If specified tests and monitoring are required, reports have to be made on these to the Minister. Similarly, compliance with a specified audit program would have to be to the satisfaction of the Minister. Fines apply for failure to comply. There is also a provision for the Minister to conduct tests and monitoring (section 48C(1b)), and presumably to recoup the costs from the developer.

Tasmania

There are no specified legislative requirements for monitoring in Tasmania, although it may be required as part of the permit conditions set by the Board of Environmental Management and Pollution Control under section 25(6)(c) of the Tasmanian Environmental Management and Pollution Control Act. The assumption is made that monitoring should occur to meet sustainability objectives under the new Resource Management and Planning System (Townsend, pers. comm., 1996). The Guidelines (DELM 1996) also outline broad requirements for enforcing compliance with permit conditions. Responsibility for ensuring compliance rests with the planning, or council, officer (level 1) or with the planning authority in conjunction with the Director of Environmental Management (level 2). Monitoring and compliance requirements are not specified for level 3 assessments.

Victoria

Monitoring is a weak link in the Victorian Act, although Wood (1993) notes that it does occur in practice. Some provision is also made in the Guidelines whereby monitoring requirements may be outlined in the assessment report, or may be undertaken by the Department of Planning and Development. Moreover, monitoring of the EIA system can be achieved with the public register of decisions established in 1988 (Thomas 1996). The Australian Intergovernmental Committee for Ecologically Sustainable Development (ICESD 1996) also notes that community monitoring committees are sometimes established to facilitate management and public accountability.

Western Australia

Monitoring in Western Australia is provided for by the decision-making authority (section 48 of the Environmental Protection Act). In the event that conditions are not being complied with, the Minister may serve an order to the proponent to either cease activities for not more than 24 hours or to comply with the conditions. It is an offence for the proponent not to respond. The EPA(WA) has also introduced an extensive 'audit program', to be prepared in consultation with the proponent. The program includes performance and compliance reports, and audit tables. Audit tables outline conditions of, and commitments to, meeting these performance and compliance reports. The reports are public documents (DEP 1996).

APPEAL RIGHTS

As noted in Chapter 1, a number of the US federal projects that had an EIS prepared under the NEPA were subject to lawsuits. In fact, litigation has been an important element in defining and refining the EIA process in the United States. During the first 13 years of the NEPA about 10 per cent of all EISs prepared by federal agencies had lawsuits filed, with a peak of 189 in 1974. Although there was a drop in the incidence of lawsuits in the late 1970s, there was still a high rate in the late 1980s (Kennedy 1986). Of the lawsuits that are filed it is estimated that almost 40 per cent are filed by environmental groups (Canter 1984), and there have been accusations that the litigation is often used as a stalling tactic. However, in a comparison of US and German EIA procedures for similar projects, Kennedy (1986) concluded that the US system actually facilitated development. This is probably related to the greater certainty in the US system: with an established history of environmental litigation, US lawsuits proceed against a background of precedents and court cases. Kennedy notes that, in contrast, West German environmental groups caused significant delays in relation to development projects subject to EIA requirements. Wathern (1988) suggests that the UK system, which has witnessed a number of protracted public inquiries with similar delaying effects, is no more reassuring.

The UK planning system does not make provision for appeal against a planning decision, although in relation to a project where there were significant potential environmental impacts, it would be possible to make a challenge on the grounds that no EIA was prepared (Glasson, Therivel, and Chadwick 1994). In the United Kingdom an important aspect of an appeal is first to establish a sufficient interest in the decision in order to gain 'standing'.

Appeal rights in Australian states and territories

In general, Australian EIA has avoided the US practice of judicial review and availability of appeal rights (see Table 5.2). In Australia the key difficulty involved in bringing an appeal is to gain 'standing' and demonstrate a legitimate interest in the proposal. Standing refers to the right to appear in court and present a case. Without this right, the public (or environmental groups) have no recourse to the courts to challenge a project on environmental grounds. However, New South Wales is unique in its generous standing and appeal rights for EIA-related development applications, and there are also provisions for limited appeal in Tasmania and Western Australia. Elsewhere there have been legal challenges to projects subject to EIA requirements, but these are often not directly linked to the EIA legislation or the adequacy of environmental investigations (as they are in the United States).

For example, in South Australia an EIS was prepared for the Wilpena Tourist Facility project (SANPWS 1988a, 1988b) but the project was

Table 5.2 Appeal rights for EIA-related projects in Australian states and territories

States and territories	*Provision for appeal rights*
ACT	Appeals made to Land and Planning Appeals Board (appeals available to those who made submissions)
	Appeals may also be made to the Administrative Appeals Tribunal
NSW	Broad 'standing' rights'
	Appeals made to Land and Environment Court
NT	No appeal provisions under the EIA process, but if proposal initially lodged under the Planning Act, appeals may be made to the Planning Tribunal
Qld	Appeals relating to decisions made under the Local Government (Planning and Environment) Act made to Planning and Environment Court
SA	No appeal rights against a decision by the Governor
Tas	Levels 1 and 2— appeal rights are available to those who made submissions
	Level 3—no appeal rights against a decision by the Governor
	Appeals made to Resources Management and Planning Appeal Tribunal
Vic	Appeal only available in respect of relevant project approval processes, but if held for EES and planning approval process, an inquiry usually provides an alternative to appeals
WA	Appeals against level of EIA or contents of assessment report can be made to the Minister for Environment
	Proponents can appeal against conditions that are part of the final approval

exempted from planning approvals under the previous *Planning Act 1982*. A legal dispute arose, not over the adequacy of the EIS but over the appropriate decision-making process exempting it from planning controls. An action brought by the Australian Conservation Council of South Australia was lost in the Supreme Court of South Australia, but the Conservation Council was given right of appeal to the High Court of Australia (*Australian Conservation Foundation Inc. and Conservation Council of South Australia v. The State of South Australia and Ophix Finance Corporation Pty Ltd* (1990) 53 SASR 349). The conservation groups managed to gain standing to present their case. However, further legal challenge was stalled, largely because the South Australian government introduced a special Act written to facilitate approvals for the tourist development, *The Wilpena Station Tourist Facility Act 1990*.

Australian Capital Territory

In the Australian Capital Territory, any person who made a submission in relation to the EIA documents and application, in relation to a particular proposal, may appeal to the Land and Planning Appeals Board established under the Land (Planning and Environment) Act (Coopers & Lybrand 1994). If an order is made against the proponent, in order to gain control of the project—such as an order to cease actions causing environmental damage—the proponent may appeal to the Administrative Appeals Tribunal.

New South Wales

The relative ease with which standing can be established in New South Wales makes appeal processes in this state unique. Broad standing rights under section 123 of the Environmental Planning and Assessment Act (Gilpin 1995) provide that anyone may appeal a possible breach in the Act via the Land and Environment Court. In New South Wales a determination that the project has environmental impacts that are 'significant' is an important part of the triggering process (Farrier 1993). Some projects may escape the 'EIA net' because the potential environmental impacts are deemed not to be 'significant'. Because 'significance' is so hard to define, this determination may be subject to dispute, and hence challenged in court. More specifically, under section 97 of the Act proponents may appeal against the consent authority's determination for Part IV decisions within 12 months of that determination. Other objectors may appeal under section 98 within 28 days of the final decision. However, no appeal is possible against matters called in under section 101.

Northern Territory

There are no provisions for third-party appeal in the Northern Territory, although if a proposal was initially lodged under the Planning Act, appeals can be made to the Planning Appeals Tribunal regarding imposed conditions. Similarly, applicants for mining permits may appeal against conditions of consent (Coopers & Lybrand 1994). In addition, Coopers & Lybrand (1994) suggest that on an informal level there are adequate negotiation processes that effectively avoid any Ministerial conflicts.

Queensland

In Queensland there is no scope for appeal under the State Development and Public Works Organisation Act, but appeal rights are present under Part 7 of the Local Government (Planning and Environment) Act, whereby both the applicant or objectors may appeal to the Planning and Environment Court (Coopers & Lybrand 1994).

South Australia

In South Australia, there is no appeal against a Governor's decision under the Development Act. However, under the previous legislation (Planning Act 1982) it was possible for registered objectors to appeal against a planning development application that was subject to an EIS—where the decision maker was not the Governor. Since the introduction of the Development Act, all avenues of appeal against EIA projects have essentially been closed off.

Tasmania

Both first- and third-party appeal rights are available in Tasmania (Townsend, pers. comm., 1996). Under section 48 of the Environmental Management and Pollution Control Act, any person who has a 'proper interest in the subject matter' may enforce EIA provisions via application for an enforcement order (Cole 1995). Although not specified in great detail in the legislation, there are four general enforcement provisions: environment protection notice, environment infringement notice, civil enforcement, and general offences enforcement. However, only those who made representations during the public exhibition period have a right to appeal. Moreover, section 28 of the State Policies and Projects Act limits appeal rights for level 3 activities: no person may take action in the event that the Governor makes an order on Ministerial recommendation. Any appeals made against decisions made by the various boards involved in the EIA process or against decisions of planning authorities are made to the Resources Management and Planning Appeals Tribunal (DELM 1996).

Victoria

The Victorian Minister's assessment report is not subject to appeal, although appeal may exist following decisions by the relevant Minister made within the normal approval processes (Coopers & Lybrand 1994). But if these approval processes are coordinated with the EIA process and an inquiry is held, as is often the case, the inquiry may become an alternative to these appeal mechanisms (Department of Infrastructure, pers. comm., 1997).

Western Australia

In Western Australia any person may, within two weeks of completion of the assessment report, appeal against the level of assessment required and against the advice in the assessment report of the EPA(WA)(Wood and Bailey 1994). Only proponents may appeal against conditions set in the final decision.

CONCLUSION

This chapter has discussed the final stages of the EIA process whereby documentation is finalised in the lead-up to decision-making. Earlier in this text, it was pointed out that the formal EIA process is completed prior to the decision-making that it is designed to inform. However, since post-decision monitoring, auditing, and appeal rights are inextricably linked to the EIA process, they have been discussed together in this chapter. Some authors think that it is difficult to separate monitoring and auditing from the process and, consequently, treat them as an integral part of EIA.

In Australia, the EIA process is designed to inform and advise. The completion of an EIS, for example, does not mean that a project is approved. It merely means that there is sufficient environmental information to inform the decision maker about the potential impacts, which should then be taken into account when making a decision. The assessment of EIA in general is usually conducted by specialist EIA branches in respective government departments, which then forward this advice to the relevant Minister. The subsequent decision-making usually rests with a responsible Minister (this is not usually an Environment Minister, except in Western Australia), but other levels of independent or internal decision-making vary between jurisdictions.

It is clear from the discussion of monitoring and auditing that some states have no formal monitoring or auditing requirements. This is an area of weakness in Australian EIA and is likely to come under greater scrutiny in the future. Similarly, mechanisms for appeal are generally weak in Australia, with the exception of New South Wales and Western Australia. This lack of appeal rights sets the Australian system apart from the US system: there has been extensive use of the courts in defining and redefining the role of EIA in the United States. However, as will be seen from Chapter 6, which examines Australian EIA in practice, there are some projects where the EIA process becomes closely linked with legal disputes.

6

EIA in practice: case studies from Australian states and territories

This chapter uses real examples to illustrate how the application of EIA legislation, regulations, and formal and informal procedures (discussed in Chapter 4) varies between states and territories. As Harvey (1995) notes, it is impossible to gain a full understanding of EIA procedures without examining how they are used in practice. Thomas goes further, suggesting that 'practitioners and academics ignore each other's interests at their peril' (Thomas 1996, p. xi).

Just as there is variation in the nature of EIA provisions between states and territories, so there is considerable variation (between states and territories) in the actual quantity of EIA documents produced at different levels of assessment (Table 6.1, p. 90). This makes the task of selecting suitable examples of EIA practice difficult, but since the EIS is a commonly accepted document around the world, the following examples use only projects at the EIS, or equivalent, level—notwithstanding the different terminology used in some states.

This chapter presents a case study from each state and territory, with the exception of the Australian Capital Territory, where the most widely used level of assessment has been the PA: there has been a total of 41 PAs since the commencement, in 1991, of the Land (Planning and Environment) Act (ACTPA, pers. comm., 1995). At the time of writing, only one EIS had been produced in the Australian Capital Territory—West Belconnen in 1992— and this was actually triggered under transitional provisions prior to the implementation of the Act (ACTPA 1992b). In addition, there have been no PERs or inquiries, and only one round-table discussion. There have, however, been a number of assessments undertaken jointly with the Commonwealth. Because only one (partially completed) EIS has been produced under Australian Capital Territory legislation, this EIS is not representative enough to be used as a case study.

The case studies from the other states and from the Northern Territory have all been selected to draw out key issues in current EIA procedures. While some of the case studies involve legislation that is no longer current, the issues involved are still relevant today. A number of the proposals selected have been controversial and have received national attention—for example, the Hindmarsh Island bridge project. Another of the case studies, Magnetic Quay, is very complex: it involved more than one jurisdiction, was subject to a number of different levels of EIA, and gave rise to various reports, to legal action, and to an independent review of the decision-making process.

The final part of the chapter comments on how various aspects of Australian EIA occur in practice—aspects such as timing of the process, public involvement, political intervention, and the use of EIA as either a decision-making tool or an implement of environmental management.

New South Wales

Statistics on current practice in New South Wales from 1989 to August 1995 are available (Department of Urban Affairs and Planning, pers. comm., 1995): a total of 1034 EISs have been exhibited (there may be some underestimates due to a change in the recording system in 1991). Gilpin (1995) states that from 1980 to 1988, 900 EISs were produced, making an approximate total of 1934 EISs since 1980, which is significantly more than other states and territories.

The Sydney Harbour Tunnel EIS

The proposal

The Sydney Harbour Tunnel was essentially a privately funded proposal to build public infrastructure, and was initiated by a joint venture between two companies, Transfield and Kumagai. This joint venture formed the Sydney Harbour Tunnel Company . The project, which evolved from two feasibility studies undertaken in 1984 and March 1986, aimed to improve traffic access across Sydney Harbour. This involved the construction of a 2.28 km tunnel comprising a four-lane road connecting Warringah Freeway in North Sydney with the Domain Tunnel in Bridge Street, near the Sydney Opera House (Figure 6.1).

The tunnel entailed sections (1300 m) to be constructed on land, on both sides of the tunnel, together with a dredged area for an immersed section of concrete units (980 m). These concrete units were to be

Table 6.1 Numbers of EIA documents produced in practice in Australian states and territories

States and territories	*Numbers of documents produced at each EIA level*
ACT	PA=41
(1991–95)	PER=0
	EIS=1
	inquiry=0
NSW	EIS=1934 (150–200 per year) (estimate)
(1980–95)	inquiry=6 (all in 1996)
NT	PER=110
(–1995)	EIS=23
	inquiry=0
Qld	EIS=180*
(1992–95)	IAS=100*
SA	DR=0
(1982–97)	PER=0
	EIS=41
Tas	No central records yet
Vic	EES=170 (10 per year) (estimate)
(1975–95)	(information as to the number of inquiries is not available)
WA	CER=543
(1987–95)	PER=139
	ERMP=20
	inquiry=0

* *These numbers are estimates only and are not directly comparable with other jurisdictions.*

pre-constructed in a casting basin to be established on reclaimed land at Botany Bay. Also planned were:

- an exhaust structure to disperse air and dilute vehicular exhaust emissions
- drainage systems
- ventilation ducts
- a four-storey administrative building
- toll plazas for south-bound traffic
- pedestrian tunnels at the north end.

The toll was set at $1.00 (indexed to inflation). It was assumed that tolls would cover tunnel costs and maintenance. Land was proposed to be leased from the Commissioner for Main Roads, Department of Main Roads, and at the end of the lease in 2022, the tunnel was to be transferred to government ownership. The Commissioner for Main Roads also agreed to loan the proponent $223 million (interest free), to be repaid to the Commissioner from tolls.

In March 1986 the then premier, Neville Wran, announced that the proposal would proceed subject to a further feasibility study, which was later completed in October 1986. A press release from the office of the premier, dated 13 March 1986, suggested that the project would 'provide a high quality harbour crossing without affecting one residence' and would 'be built at half the cost of previous schemes and more quickly' (McNamara 1986, p. 2).

The proposal originally fell within the scope of both Part IV and Part V of the Environmental Planning and Assessment Act, and an EIS was formally required in June 1986. Under the terms of Part V of the Act, the Commissioner for Main Roads in the Department of Main Roads was the nominated determining authority. However, determination could only be made if the Minister for Roads was satisfied that Part V of the Act had been complied with.

Under the terms of Part IV, Randwick Council noted that the proposed concrete batching plant that was to be used to construct the immersed concrete units was a designated development under Schedule 3, and would therefore require a formal development application, including an EIS. However, legislation was later introduced (*Sydney Harbour Tunnel (Private Joint Venture) Act 1987*—assented to 28 May 1987) to restrict the EIA process to Part V of the Act so that control remained with the Commissioner for Main Roads.

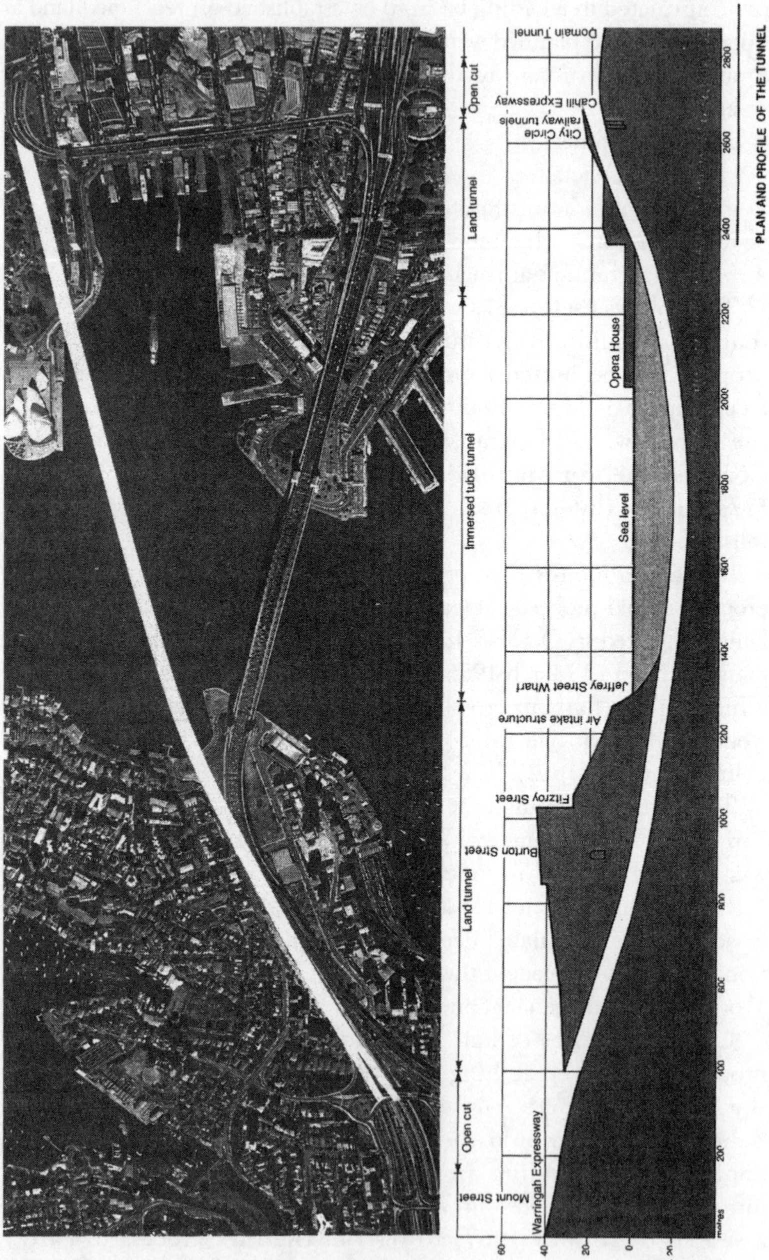

Figure 6.1 Sydney Harbour Tunnel proposal

Source: McNamara 1986.

The EIS

The Director of the Department of Environment and Planning was requested, on 2 June 1986, to determine the form and content of the EIS (for the whole development), as required under regulation 58 of the Environmental Planning and Assessment Act. A number of government agencies were consulted (these agencies are listed in an appendix to the EIS), and the information was subsequently produced, within the required 28 days, on 30 June 1986 (note that no public comment is required under the Act).

The EIS was completed in November 1986 (McNamara 1986), around the same time as the feasibility study, and contained:

- a traffic and transportation assessment
- a consideration of alternatives, all of which were rejected by the proponent (for example, the 'no go' option, alternative bridges, new bus lanes, rail tunnels, light-rail system, alternative tunnel alignments)
- a justification of the proposal using cost–benefit analysis
- design options
- the environmental context
- an assessment of environmental impacts and the safeguards necessary to prevent them occurring
- an energy statement.

A number of potential environmental issues were identified, but most were not considered to be significant.

Potential environmental issues in the Sydney Harbour Tunnel EIS

- land-use, community, and visual impacts
- adverse air quality from carbon monoxide emissions
- heritage impacts
- noise impacts from construction such as dredging, drilling, and blasting
- regional land-use-planning impacts (emphasised as positive)
- impacts from casting basin, such as effects on water quality, impact of de-watering on adjacent sites, visual impacts (tower cranes), air quality (sand), noise, and truck movements
- other impacts (traffic, utilities, marine ecology, and water quality) deemed insignificant

Overall, it was concluded that:

> the proposed Tunnel is justified on travel, economic, energy and air quality grounds. The Tunnel will create environmental benefits as a result of traffic reduction in the city centre, improved cross-Harbour access, energy savings, air quality improvements, emergency response improvements and increased scheduling reliability for buses. Although there will be adverse environmental impacts . . . they will be minimised by the application of the environmental safeguards . . .
>
> McNamara 1986, p. 15

Following consultations with the Department of Main Roads, and before public review, a later addendum regarding minor refinements to the proposal was incorporated into the EIS.

Public review

Public review of the EIS extended beyond the minimum requirement of 30 days, with an initial period from 19 December 1986 to 20 January 1987. Further information regarding finance was displayed on 12 February 1987, and additional submissions were received by the Department of Main Roads until 9 March 1987. Unlike the EIA process in most jurisdictions, in New South Wales there is no legislative requirement for the proponent to respond to the public submissions in a supplement or final EIS. The key issues were later summarised by the Department of Environment and Planning (DEP 1987) and by the Department of Main Roads (DMR 1987). The latter was required by the Act to consider any representations in making the final decision.

Key issues of public concern in relation to the Sydney Harbour Tunnel EIS

THE EIS PROCESS

- inadequate public exhibition period (over Christmas break)
- inadequate EIS (biased, lacked thoroughness, misleading)
- failure to objectively consider alternatives
- broader context of comprehensive metropolitan transportation plan required

- failure to consider regional planning issues
- further consideration and further studies needed
- given the proposal's significant nature, a commission of inquiry should have been undertaken (DMR 1987)

THE PROPOSAL

- plan would cause increase in traffic and actually exacerbate congestion on the Bridge and local areas
- criticised on the basis that a tunnel does not encourage public transport
- aggravation of air pollution
- toll deemed an 'unprecedented selective tax' on north shore residents
- impacts of traffic flows on local quality of life
- loss of Bradfield Park as an important local and regional open space for long period
- failure of government to open proposal to public tender
- unclear financing arrangements

There was disagreement between the departments over the nature and extent of public concern, and this disagreement was, in part, determined by the way in which the departments had summarised the submissions.

The Department of Environment and Planning noted that public concern was significant both during and after the public review period. It noted that before the review period 443 submissions were received: 24 from government departments, 9 from local councils, and 410 from private sources (together with four petitions containing, in all, 250 signatures). In contrast the Department of Main Roads noted 463 submissions: 25 from government departments, 9 from councils, and 429 from private sources.

According to the then Department of Environment and Planning, 84 per cent of submissions opposed the proposal (including the petitions), 4 per cent voiced support (particularly because of the potential reduction in heavy traffic congestion and the time and energy savings), and 12 per cent made general comments.

Of particular interest was the submission from the Minister of Transport (on behalf of the Traffic Authority, the Urban Transit Authority, and the State Rail Authority) noting the inadequacy of the EIS in terms of its failure to assess a number of issues including:

- wider traffic implications and associated effects in the central business district
- the need for reassessment of road capacity and management
- the costs of ancillary works to maintain traffic movement efficiency
- the impacts on bus routes and times
- the potential for increased pressure for commercial development on the lower north shore (DEP 1987).

In contrast, the Department of Main Roads argued that only 64 per cent opposed the proposal (petitions excluded), 9 per cent supported the project, and 27 per cent made comments. They further noted that 141 of the submissions were in the form of two standard letters, and that the Department of Environment and Planning was 'arbitrarily selective in highlighting concerns of key government authorities' and was not entirely accurate when summarising council concerns (DMR 1987, Appendix E2, 3). They further noted that

> The Ministry of Transport predicates its response on behalf of the Transport Administration with 'the Transport Administration has no fundamental objection to enhancing cross-harbour capacity so long as the consequent impacts are adequately addressed'. Many of the 'concerns' hinge on methodology. During construction the Ministry sees 'none of the traffic impacts as being of sufficient magnitude to justify not proceeding with the project . . . if adequate machinery is established to identify and implement the necessary solutions'.
>
> DMR 1987, E3.

The independent director's examination

Pursuant to section 113(3) of the Environment Planning and Assessment Act, submissions were received by the Department of Main Roads and forwarded to the Secretary of the Department of Environment and Planning to determine the need for either an independent examination by the Director General (section 113(5)) or a commission of inquiry (section 114). In this case, the former was required by the Director, and the Department of Main Roads was notified on 4 March 1987, before completion of the extended public review period. This decision was made in light of several technical problems, and contradictions, in the EIS.

A multi-disciplinary team was established, independent advice was sought regarding traffic, economics, and the casting basin, and the proponent forwarded additional information to aid in the examination. The final report, prepared by the Department of Environment and Planning for the Director, was completed on 27 March 1987, within 24 days of receiving the EIS. It was highly critical of the EIS, suggesting that

> the information and analysis contained in the Environmental Impact Statement and supporting documents is deficient and in some cases misleading . . . the [EIS] is in reality a 'sales' document which fails to provide justification for the construction of the Tunnel. The Environmental Impact Statement does not provide an adequate evaluation of the project and seriously overstates the benefits and understates the costs to the community and the risks to the Government.
>
> DEP 1987, p. 1

More specific issues raised were similar to those raised in the public review period.

Issues raised in the independent director's examination of the Sydney Harbour Tunnel EIS

- incompatibility with regional planning objectives
- potential fall in public transport use
- lack of quantitative data about the impacts of tunnel construction (for example, noise impact, dirt, air pollution, parking loss, loss of recreational and visual amenity, valuable tree loss, water pollution, traffic impacts from trucking)
- inadequate consideration of risks from Port Botany casting basin (for example, impacts on soil foundations, and the pollution of the Port Botany aquifer)
- unsound traffic projections (overestimated)
- near saturation of arterial roads (capacity overestimation)
- a bias in the presentation of benefits to peak hour commuters and developers
- proposal not viable in economic terms (overestimation of benefits and underestimation of costs)
- overstatement of fuel savings
- need to further evaluate alternatives

The report also stressed the importance of the potential risks to the Government, suggesting that the Commissioner for Main Roads could conceivably be forced to continue subsidising the project if traffic, and hence toll collections, turned out to be less than predicted. The report noted that the government could gradually incur a net loss of $87 million (1987 values). Overall, the report recommended that the project should not proceed, and that viable alternatives should be fully considered before a final decision was made.

The public assessment report

In making the final decision, the Commissioner for Main Roads was required to 'consider the report [of the Director's examination]' and to incorporate this consideration, together with an examination of the EIS and any representations made by the public, into a public assessment report (regulation 64 of the Environmental Planning and Assessment Act). It is interesting to note that the Commissioner of Main Roads is often both the proponent and the determining authority (DMR 1987), and this raises some concerns regarding potential bias in the assessment process and decision. However, in this case, because the Department of Main Roads was not the proponent, the Department of Main Roads concluded that the scope of their assessment report was an *independent* examination of representations made on the EIS and all other documents.

The assessment report was completed on 6 April 1987 (DMR 1987), only six working days after completion (by the Department of Environment and Planning) of the report on the Director's examination. There appears to be some disparity between the conclusions drawn in the report on the Director's examination and the conclusions drawn in the assessment report of the Department of Main Roads. The brief time-frame suggests that the recommendations of the report on the Director's examination—to conduct further 'detailed' investigations—were not fully pursued. However, under the legislation (regulation 64 of the Environmental Planning and Assessment Act), the determining authority need only *consider* the Director's examination, thus highlighting the discretionary and subjective nature of the process. In addition, two weeks prior to the independent Director's examination, the Department of Main Roads had sent a letter to the proponent outlining a number of draft conditions of approval that would appear in the Commissioner's determination. This tends to raise some questions about the Director's influence in practice.

Nevertheless, in formulating the assessment report, the Department of Main Roads did request further information from the proponent. The Department also undertook further consultations with government departments and sought additional expert advice. Alternative proposals were reviewed, as recommended in the report arising out of the Director's examination, and it was again concluded that these alternative proposals were either inferior to the Harbour Tunnel or were more costly with greater environmental impacts.

When considering the submissions received during the public review period, the assessment report essentially justified the proposal in terms of traffic need and associated benefits, and tended to agree with the original EIS findings. The assessment report found that if conditions of mitigation were met, many of the construction and operational issues raised could be characterised as minor and surmountable. The assessment report considered, but did not respond to, issues raised by the public, regarding the preparation of the EIS—for example, consistency with the Director's requirements, need for further studies, bias, and public review period. Finally, pursuant to section 112(4) of the Environmental Planning and Assessment Act, the assessment report outlined a number of conditions (for example, in relation to noise and blasting, hours of construction operation, land use, and community issues, spoil disposal, restoration, utilities, and traffic issues) and project modifications for consideration by the Commissioner for Main Roads when making the final determination. Further investigations were also requested—for example, in relation to the casting basin and heritage impacts.

The final determination

The final determination was made by the Commissioner for Main Roads on 29 May 1987: the proposal was approved. In order to facilitate the 'construction, maintenance and operation' of the tunnel, the Sydney Harbour Tunnel (Private Joint Venture) Act was passed and subsequently amended by the *Sydney Harbour Tunnel (Private Joint Venture) Amendment Act 1987* (16 December). The Sydney Harbour Tunnel (Private Joint Venture) Act not only restricted the assessment process to Part V of the Environmental Planning and Assessment Act, as noted earlier, but it also prevented any appeals in relation to:

- the Commissioner's decision
- any conditions set under section 112 (where an EIS is required)
- any failure to comply with conditions set under section 112
- any decision under section 111 (duty of determining authorities to consider the environmental impacts of activities)
- any failure to comply with the EIA provisions under Part V.

Thus this Act reduced accountability. This is significant in that New South Wales is noted for its broad avenues for appeal in the EIA process relative to most other jurisdictions.

The fallout

Significant controversy followed this approval, and prior to state elections in 1988, the Liberal Opposition promised to hold an inquiry into the tunnel and pass a stop-work order. The newly elected Liberal premier, Nick Greiner, suggested to the media that 'building the Sydney Harbour tunnel was not the right decision to solve Sydney's traffic problem . . . If it had been properly assessed, it would get a pass mark, but it wouldn't be the best option' (Coultan 1988, p. 2). Yet no inquiry was held, as previously promised by the Liberals. This was not surprising given that special legislation would have been required to stop work (ENERSOL 1989). Consequently, the North Sydney Municipal Council held their own inquiry into the tunnel. They did not, however, have access to confidential documents such as the contracts, and the only information made available was that which was already publicly accessible (ENERSOL 1989).

Nevertheless, the Council's public inquiry was a detailed study. It was highly critical of the EIA and the decision-making process. The inquiry noted substantial errors in the EIS that 'boosted the benefits' of the tunnel with, among other errors, an inaccurate cost–benefit analysis, limited consideration of alternatives, inadequate consideration of the physical impacts, tunnel design problems, and an overemphasis on car usage. A total of 74 submissions were also received during the inquiry, 61 of which opposed the tunnel and 4 of which supported it. The issues that were predominant in the submissions related to traffic and transport, followed by environmental effects and procedural problems. As a result of the inquiry, the North Sydney Municipal Council made six recommendations, including, among other things, that:

- work be stopped and a state inquiry into the process be initiated
- if work could not be prevented, two lanes be dedicated to public transport
- future proposals of this nature should not be allowed to bypass appeal rights.

However, no state inquiry was held and the tunnel was completed in 1992. Since then it has drawn significant attention in the media, particularly in relation to controversial financial arrangements and an 'ensured revenue stream' agreement outlined in the Sydney Harbour

Tunnel (Private Joint Venture) Amendment Act. Money collected by the government from tolls did not cover the money loaned by the government to the proponent, resulting in a shortfall of $3 million as of 1994 (Cook and Voumard 1994). Predictions of losses arising from the tunnel for both government and taxpayers varied, ranging from $650 million to $5.2 billion. This issue of financial losses to the government is important given that the Director's independent examination had, in fact, predicted financial losses for the government. It also highlights the advantages of the recent amendments to the Environmental Planning and Assessment Act specifying that, where the determining authority is also the proponent, a proposal under Part V cannot proceed without the consent of the Minister for Planning. Despite the fact that the determining authority (Commissioner) was not the proponent in this case, the amendments are useful in similar situations where the proponent/determining authority has a strong interest in the proposal and appears to ignore public and government comments raised during the EIA process.

Northern Territory

In the Northern Territory the majority of assessments have been in the form of PERs, with approximately 110 (DLPE, pers. comm., 1995) and approximately 23 EISs. Given that the PER is a recent introduction, the official figures probably refer to the Preliminary Environment Report (PrER), which does not require public review, rather than the Public Environment Report (PER). The case study selected involves both Northern Territory and Commonwealth EIA legislation.

The Tindal airbase EIS

The proposal

In 1982, following the consideration of alternative sites by the Commonwealth Department of Defence, the existing Tindal Royal Australian Air Force (RAAF) base in the Northern Territory was chosen for potential upgrade to a 'home base', in order to accommodate a squadron of new F/A-18 fighter aircraft and personnel. A two-stage development providing facilities for home basing was proposed (Figure 6.2) as a replacement for existing temporary structures (construction to commence mid 1984). Also proposed were new

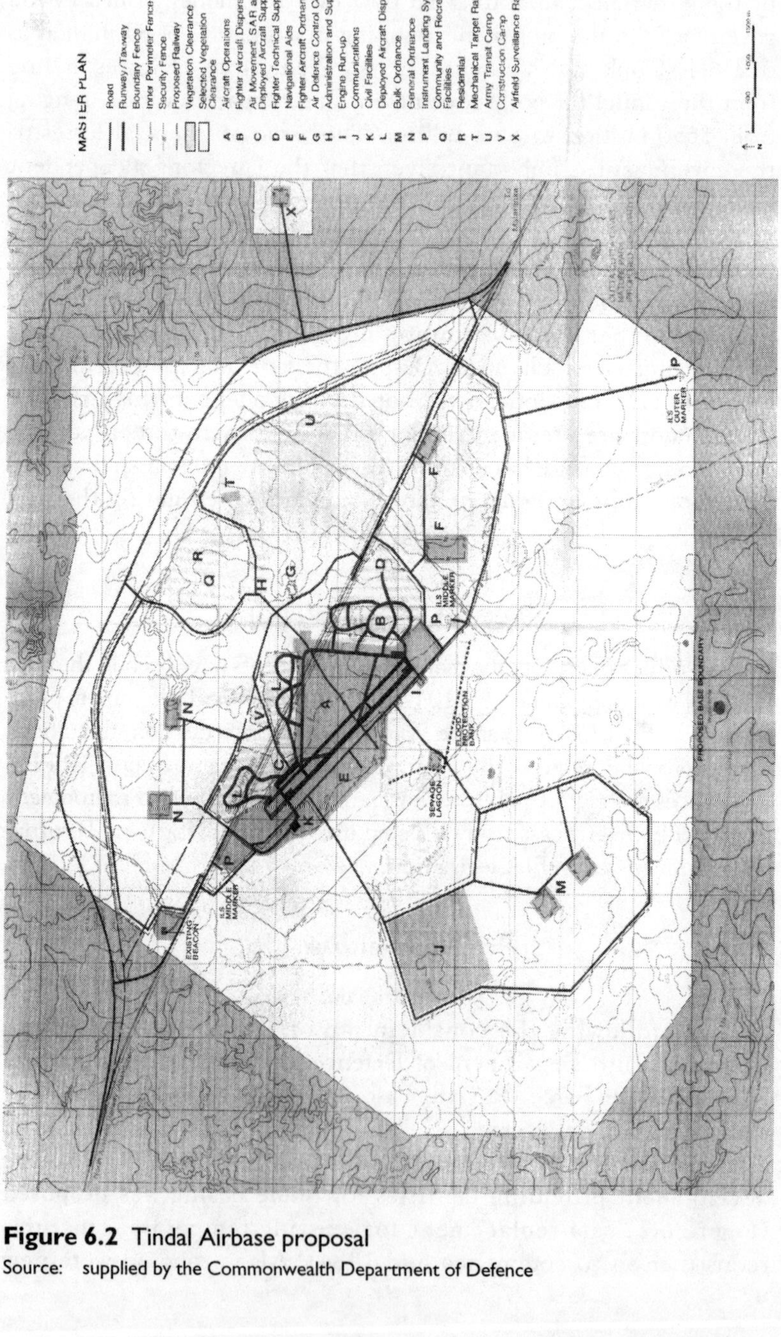

Figure 6.2 Tindal Airbase proposal

Source: supplied by the Commonwealth Department of Defence

facilities to cater for functions originally undertaken at the RAAF Base in Darwin (construction to be completed by 2000). The completed works were to include accommodation for 1000 personnel and nearly 200 civilians.

Specific components of the proposal included:

- expansion of the existing base boundary (from 45 km^2 to 122 km^2)
- an extended runway and associated taxiways and exits
- a fire station
- search and rescue facilities
- a helipad
- various other constructions to support zoning requirements—for example: aircraft dispersal zones; ordnance storage; fighter technical support; navigational aids; a control centre; facilities for administration, support, and community services; and facilities for community, recreational, and residential purposes.

Off-base facilities were also proposed, including a weapons range (2200 km^2), a high frequency remote receiving station (40 km^2 site), off-base housing in Katherine, radar, and beacons. In order to facilitate the proposal, a second transmission line from Katherine was proposed for electricity (this transmission line was the subject of a separate Northern Territory EIS). Realignment of both the Stuart Highway (this realignment was the subject of a separate Commonwealth EIS) and the proposed Alice Springs-to-Darwin railway was also required.

The requirement for an EIS formed part of a broader feasibility and planning study consisting of a master planning process, an ongoing design and development process, and a decision process operating simultaneously with the EIS process. In terms of the EIA process, the fact that the proposal was a Commonwealth one triggered the Commonwealth *Environment Protection (Impact of Proposals) Act 1974*. However, because the proposal also involved the Northern Territory government (for example, in infrastructure), a joint assessment agreement was confirmed in July 1983, whereby both the then Conservation Commission of the Northern Territory (CCNT) and the Commonwealth Department of Home Affairs and Environment would assess the EIS and the supplement. Compliance with the Commonwealth procedures was considered sufficient to satisfy Northern Territory requirements, which is not surprising given that the Northern Territory process was essentially modelled on the Commonwealth's. The joint process was facilitated with the establishment of a Joint Consultative Committee consisting of representatives from the governments of the Commonwealth and the Northern Territory, from the Katherine Town Council, and a RAAF representative.

In accordance with paragraph 2.1 of the Commonwealth Administrative Procedures, established under the Environment Protection (Impact of Proposals) Act, the Department of Defence provided preliminary information to the Department of Home Affairs and the Environment on 23 December 1982. Subsequently, the Minister for Home Affairs and Environment required an EIS on 13 April 1983, although CCNT (1984) notes that the Department of Defence was actually notified in March. Because these events took place prior to the 1985 amendments, the legislation did not provide for PERs. The Department of Defence was nominated as the proponent and the Department of Housing and Construction had a major role in construction operations associated with the proposal.

The EIS guidelines and the draft EIS

Guidelines for the EIS were jointly prepared in March 1983 (CCNT 1984), although the official EIS requirement came later in April. There does not appear to have been any public input. However, it should be noted that at the time of the proposal public input was discretionary under the Commonwealth process and was not required under Northern Territory processes (Coopers & Lybrand 1994)—this changed in the Northern Territory in 1994 when public review provisions were introduced for the EIS Guidelines (these involve a 14-day review period).

The draft EIS was subsequently completed in November 1983 by consultants Kinhill Stearns on behalf of the Department of Defence (Kinhill Stearns 1983). The draft EIS established a brief appraisal of:

- the need for the proposal
- development and planning requirements—for example, zoning options
- the existing environment (including social factors)
- effects of the proposal, including socio-economic and physical effects
- preferred sites for the weapons range
- environmental management techniques to mitigate effects.

Key issues arising from both construction and operation of the proposed development are listed in the box below.

Key issues in the Tindal Airbase Draft EIS

SOCIAL, ECONOMIC, AND DEMOGRAPHIC EFFECTS

- population change, with an increase of 78 per cent in the first stage
- pressure on existing community facilities and services in the town of Katherine
- pressure on existing housing and accommodation
- alterations in existing social networks and attitudes
- modifications in existing community and decision-making structure
- increases in local employment and town revenues
- impacts on the Aboriginal community: decrease in the proportion of Aboriginals in relation to the whole population, and associated problems of becoming a minority in terms of health, morale, services, and access to planning and decision processes
- effects of aircraft noise
- increase in economic activity (in terms of output and employment) in the Northern Territory and Katherine
- impacts of land acquisition

PHYSICAL EFFECTS

- impacts on regional land use (for example, recreational demands on natural tourist attractions such as the Katherine Gorge National Park, or effects on pastoralists)
- increased water requirements and supply
- increased demand on sewerage system
- impacts arising from the sourcing and transportation of construction materials (for example, the requirement for 500 000 m^3 of crushed rock, and 200 000 m^3 of sand)
- increase in traffic density
- air emissions from aircraft and associated operations
- increased waste-disposal requirements and toxic wastes from the base (for example, fuels)
- aircraft noise and effects on building design
- potential impacts on hydrology, ecology (via land clearance), fauna, and wetlands
- potential for bird strike
- impacts on Aboriginal or archaeological sites

Kinhill Stearns 1983

Public review, the public awareness program, and public hearings
Although not required by legislation, preliminary drafts of the EIS were reviewed, prior to public review, by both CCNT and the Department of Home Affairs and the Environment. The EIS was subsequently released for public inspection, for six weeks from 7 November until 19 December 1983; the last submission was received on 13 January 1984. This period clearly extended beyond the minimum requirement of 28 days for both the Commonwealth and Northern Territory. The number of public submissions was relatively low (28), and the majority of them were from government sources (16 from the Northern Territory government, seven from the Commonwealth). Five were from private sources.

As required by both Commonwealth and Northern Territory legislation, the public issues were responded to by Kinhill in a supplement document, which formed the final EIS. This was released on 12 March 1984 and was received by the Department of Home Affairs and the Environment on 23 March. Key concerns were outlined, further information was requested, and suggestions for management were outlined. Other comments tended to agree with, and noted the accuracy of, the draft EIS.

Key issues raised in the public review phase, as identified in the Tindal Airbase supplement document

- inadequate information on the proposal provided in the public review period
- social impacts (for example, impacts on the Aboriginal community, land acquisition, population increase, impacts on community facilities, noise)
- economic impacts (for example, effects on agricultural and business sector, the decommissioning of the Darwin base)
- hydrological impacts
- problems in the EIS itself (for example, inadequate consideration of cumulative or indirect effects, no evaluation of the 'no go' option or of alternative locations)
- impacts on the natural environment (for example, potential contamination of Tindal Aquifer and Four Mile Creek from waste effluent, impacts on fauna, vegetation clearance, and soil erosion)
- impacts of weapons range (on the natural environment, on archaeological sites, on culture, and on land use)
- unclear management arrangements (for example, in relation to the weapons range)

Following identification (during the preparation of the draft EIS) of a lack of public knowledge about the proposal, the proponent undertook a public awareness program that ran simultaneously with the formal public review period. The program included:

- media releases
- a public display in Katherine and Darwin
- a public meeting in Katherine
- a household survey of Katherine residents
- a series of group discussions with interest groups
- an Aboriginal consultation program
- a survey of RAAF personnel and their dependants.

The issues raised were similar to those in the submissions, with the majority focusing on adverse social effects arising from the increased population. Notwithstanding these issues, many of those participating in the public awareness program also acknowledged the overall benefits of the proposal—for example, increased support in terms of community services, new groups and interests involved in the community, more health services, and increased employment. However, those involved in the Aboriginal consultation program had negative attitudes towards both the proposal and the consultation process. Shortly after the completion of the draft EIS, the proposal design was revised and refined in a master plan with an outline provided in the supplement, but no real opportunity for the public to comment.

On 15 December 1983, during the public review period a concurrent decision-making process was initiated whereby the Commonwealth Senate referred the proposal to the Parliamentary Standing Committee on Public Works for consideration under the *Public Works Committee Act 1969*. As part of this process, public hearings were also conducted in Katherine on 21 and 22 March 1984, and written submissions were received from government and private sources. Again, social effects were the major issues raised and addressed, with an emphasis on Aboriginal concerns. A meeting was held with Aboriginal representatives on 29 March, resulting in ongoing representation of two Aboriginals on the Joint Consultative Committee.

The assessment report

Following the public review period, the Department of Home Affairs and the Environment was required, under the Commonwealth EIA legislation, to prepare an assessment report considering the final EIS within 42 days. Following further consultation with government agencies, this was completed in May 1984 (DHAE 1984), within the required time frame. In accordance with the joint assessment arrangements and the Northern Territory Environmental Assessment Act, CCNT also prepared an assessment report in June 1984 (CCNT 1984), shortly after the completion of the report by the Department of Home Affairs and the Environment. No disparity was evident between the jurisdictions. Social effects were again noted as significant in both reports, particularly the effects on the Aboriginal community, but both reports found that it was possible to minimise these with proper management, consultation, and further studies.

The inadequacies of information within the EIS were also highlighted in the Commonwealth report, but were attributed to the conceptual nature of the proposal and the ongoing design process. The Commonwealth report also identified issues needing further consideration. These included:

- water supply sources
- electricity supply
- accommodation (location and type)
- ratio of on-base to off-base housing
- survey of the Aboriginal communities regarding the Stuart Highway realignment
- source of the construction materials to be used
- timing and funding of community services and facilities.

It was, however, stated that these deficiencies were not significant enough to prevent proper assessment in general terms. It was also noted by both reports that some issues receiving cursory treatment (for example, infrastructure expansion) were outside the jurisdiction of the proposal and would, in the event that Commonwealth funding was provided, be examined separately and in accordance with the Northern Territory EIA process and the Commonwealth Environment Protection (Impact of Proposals) Act. This raises questions about the cumulative nature of the proposal, in particular about the potentially fragmenting approach to the decision-marking process—for example, the proposed Alice Springs-to-Darwin railway and the Stuart Highway were also undergoing separate EISs.

Despite these problems the assessment report by the Department of Home Affairs and the Environment stated that the objects of the Commonwealth Act had been met and that environmental effects had 'been identified and examined as fully as is practicable at the current stage of planning' (DHAE 1984, p. v). In accordance with paragraph 9.3 of the Administrative Procedures, 24 environmental recommendations were made relating to financing arrangements, the ongoing and expanded role of the Joint Consultative Committee (with environmental and Aboriginal representation), employment, Aboriginal sites, the design and construction phases, and monitoring. Sixteen recommendations were also outlined by CCNT in relation to:

- the Tindal aquifer
- weapons range management plans
- further assessment of the Remote Receiving Station Site at the preliminary environmental report level (this level no longer exists in the current legislation)
- further studies on noise and its effects on fauna
- design of flood-mitigation levee banks
- an expanded role for the Joint Consultative Committee
- provision for the reception and due consideration of noise complaints
- proper Aboriginal representation in the proposal.

It was concluded overall that

> The adoption of the recommendations from this and the Commonwealth Assessment Report will result in the economic and social benefits of the proposal outweighing its detrimental environmental impacts.
>
> CCNT 1984, p. 3

The final decision

Under both the Commonwealth and Northern Territory processes, it is usually the action Minister or the responsible Minister who, after considering the final EIS and the recommendations in the assessment report, makes the final decision. In this case the Minister's decision appeared to be based more on the report prepared by the Parliamentary Standing Committee on Public Works, which was completed on 3 May 1984, and subsequently presented to Parliament on 29 May 1984, than on the assessment report prepared by the Department of Home Affairs and the Environment. Given that only the final EIS was presented as evidence to the Standing Committee, it is unclear how the assessment reports were considered in the decision-making process.

The Minister's report generally recommended that the proposal proceed, and suggested that the EIS recommendations to mitigate effects be adopted, together with increased and more effective consultation with Aboriginal groups. The proposal was endorsed by Parliament on 7 June 1984 and was finally approved by Cabinet in August 1984.

QUEENSLAND

Given that EIA in Queensland is fragmented, with different requirements outlined in various Acts (see Chapter 2), there is no centralised body to maintain records on the numbers and types of IASs or EISs undertaken, and it is therefore difficult to compare the practice of EIA in Queensland with its practice in other jurisdictions. There are, however, some preliminary numbers for processes undertaken under the Local Government (Planning and Environment) Act from 1992 to 1995 (Department of Housing, Local Government and Planning, pers. comm., 1995): in this period a total of 180 EISs were undertaken for works relating to Council policy (for example, residential development, water, canals), chemical processing or storage, bitumen works, aquaculture, abattoirs, concrete batching, excavation or extractive works, floodline development, major shopping developments, tourist developments, and waste treatment among others. Less than 100 IASs were undertaken through the State Development and Public Works Organisation Act by various government agencies during 1993–94 (DPETD 1995).

The Magnetic Quay IAS, PERs, and EISs

The proposals

Numerous developments have been proposed for Nelly Bay on Magnetic Island, beginning in 1971 with approval for a hotel, a breakwater, and a marina. Work was commenced, but ceased in 1974 when the land was sold with the aim of constructing a larger development, which also did not proceed. A development for home units and buildings was later proposed and approved in 1983, but again did not commence. Soon after, the Magnetic Keys development was proposed and this proposal triggered the EIA process in 1985 . This was followed by a revised project in 1988 that attracted further EIA documentation and then another revised project in 1995 that attracted even more EIA documentation. These two revisions and subsequent documentation are discussed later in this case study.

Magnetic Quay (formerly Magnetic Keys) EIA requirements

- 1985—Queensland Land Administration Commission (the responsible authority) requires an IAS

- 1985—draft IAS (deemed insufficient as basis for approval)
- May 1986—revised IAS completed
- 1988—PER required under the Commonwealth Environment Protection (Impact of Proposals) Act
    - initial PER deemed inadequate for assessment
    - second PER required
- 1992—independent review of the decision-making process required
- 1995—EIS required under both Queensland and Commonwealth EIA procedures

The initial proposal for a tourist resort development entitled 'Magnetic Keys' involved the construction of a complex on the north-east section of Nelly Bay. This complex was to include breakwaters and reclamation; a dredged basin and channel; upgrading of the power supply, communications, and sewerage; and water supply services. The key elements of the proposed development comprised:

- a tavern
- a motel complex (50 units)
- a hotel
- a ferry terminal (in the form of a floating pontoon)
- a shopping centre
- a hire complex
- offices
- a barge terminal
- service industry facilities
- a kiosk
- a marina and facilities (150 floating berths)
- a charter boat operator
- a water-front restaurant
- townhouse accommodation
- a resort hotel or motel
- a nature park and associated public facilities
- roads
- car parks

Triggering and the IAS

In September 1984 discussions between the Federal–State Offshore Developments Committee and the Queensland government determined that an EIA was necessary for the initial proposal, with the Great Barrier Reef Marine Park Authority (the Marine Park Authority) being a party to the assessment. An IAS was thus formally required by the responsible authority (the Queensland Land Administration Commission) under section 29 of the State Development and Public Works Organisation Act. In April 1984, the Marine Park Authority and Townsville City Council gave 'approval in principle' for Bayvista Proprietary Limited to be granted a special three-year lease to undertake the environmental assessment. Based on this approval, the lease was granted in 1 July 1985. This lease allowed Bayvista to access the site in order to conduct studies related to the EIA.

An intergovernmental working group was also established in April 1986, their job being to coordinate the approval process. This group included representatives from the Land Administration Commission (the responsible authority), the Marine Park Authority, the Department of Harbours and Marine, the Department of Mines, the Townsville Harbour Board, and the developers and their consultants. Townsville City Council was also included.

The draft IAS, completed only a month later (July 1985) by Bayvista Proprietary Limited, did not address the concerns of the Marine Park Authority. Later, the Marine Park Authority and the Land Administration Commission determined it to be insufficient as a basis for approval. Consequently, in November 1985, following consultation with the Marine Park Authority, the Planning Branch of the Department of Lands produced terms of reference, and guidelines, for a revised IAS. There does not appear to have been any public involvement at this stage, although the requirement for public involvement is discretionary in the Queensland EIA process.

In May 1986, following ongoing consultation with local and state governments, the consultants McIntyre & Associates completed an IAS (two volumes, the latter including technical reports) on behalf of North Barrier Holdings Limited (formerly Bayvista) (NBH 1986). The IAS was released for public comment from 12 May to an unspecified date (June–August), and submissions were forwarded to the Queensland Lands Administration Commission (the responsible authority). Again,

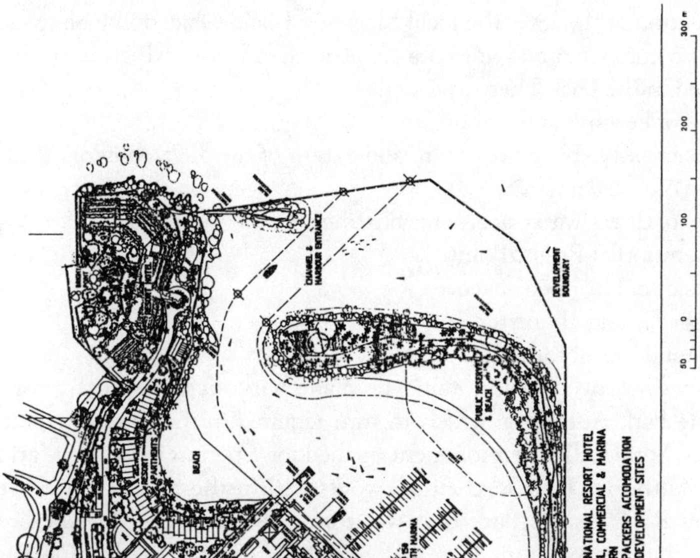

Figure 6.3 Magnetic Quay proposal
Source: supplied by McIntyre & Associates Pty Ltd

this process of allowing public consultation is discretionary under the State Development and Public Works Organisation Act. During the preparation of the IAS, the local Magnetic Island Objection Committee was also contacted and voiced a number of concerns, which were summarised in the IAS. These included:

- loss of beach and foreshore
- uncertainty about the nature and extent of the development
- loss of visual amenity
- loss of direct water-access for boating
- alteration of Bright Point
- loss of public open space
- loss of island character
- general loss of lifestyle.

However, no decision could be made without a permit from the Marine Park Authority, which in turn required further impact studies. On 13 April 1987, the proponent applied for a permit from the Marine Park Authority for construction works within the part of the Great Barrier Reef Marine Park that was zoned 'General Use A' (the area below the mean low-mark in Nelly Bay). On 5 May the Marine Park Authority defined the proposal as environmentally significant and the Minister responsible for the Marine Park Authority triggered the Environment Protection (Impact of Proposals) Act by nominating North Barrier Holdings Limited (renamed as Magnetic Keys Limited) as the proponent. On 20 May 1987, in order to determine the need for assessment, the then Commonwealth Department of Arts, Heritage and Environment requested further information from Magnetic Keys Limited. Magnetic Keys Limited forwarded this information to the Department almost a year later on 15 February 1988 (Whitehouse 1992).

Changes to the proposal

In 1988 the development proposal, now named 'Magnetic Quay', had changed slightly in terms of layout and composition, although it was still a major tourism and commercial development (Figure 6.3). The key objective was to establish an all-weather port and harbour, with a marina and ferry terminal, together with accommodation, commercial, and resort facilities. The key elements included:

- 600 m of breakwaters
- 15 ha of reclamation and dredging (12.3 ha marina basin)
- a resort hotel

- a centre for retail, commerce, and marina activities (150-berth marina, and barge and ferry terminal)
- a tavern
- backpackers accommodation
- unit development (between 74 and 94 units spread out over two sites)
- public boardwalks
- a trailer park
- a boat ramp
- a public reserve
- two new beaches (one for public use).

The PER

Under the new 1987 amendments to the Environment Protection (Impact of Proposals) Act, a PER was required on 5 April 1988 from the Minister of Arts, Heritage and Environment. This was the first PER required under the amended Act. There was, however, some later criticism that the development should have triggered an EIS (CEPA, pers. comm., 1995). It is also interesting to note that the potential for triggering the Commonwealth EIA process was known in October 1986, prior to the preparation of terms of reference for the previous IAS. Given this knowledge it is odd that, rather than requiring both an IAS *and* a PER (as was the case), a joint assessment process was not undertaken to minimise duplication. Although, it is true that at this time joint assessment arrangements were not as established as they are now.

Following consultation with the Marine Park Authority and the proponent, guidelines for the PER were produced by the Commonwealth Department of Arts, Sport, the Environment, Tourism and Territories (DASETT) on 27 April 1988 (DASETT replaced the Department of Home Affairs and the Environment). Based on advice received from the Attorney-General on 2 June 1986 and on 31 August 1988, the guidelines restricted the scope of the environmental assessment relative to the IAS. In this case, the Marine Park Authority, when assessing the permit application, could only have regard to development inside the Marine Park—that is, the marina basin, the island marina-edge treatments, and the access channel. All other issues, such as the tavern, resort developments, and other similar components of the development, lay outside the scope of the jurisdiction of the Marine Park Authority.

Public review of the PER

In May 1988, the PER was completed by consultants McIntyre & Associates, on behalf of the proponent (MKL 1988a). Because the PER essentially built on information from the IAS (Whitehouse 1992), the PER only took approximately six weeks to prepare. Prior to release, the proponent requested that the PER be exempt from public review. The department advised against this request, and the PER was released for public exhibition from 18 May to 15 June 1988, although DASETT (1988) suggests the period extended to 25 June 1988.

A total of 56 submissions were received by DASETT and were summarised in the assessment report (DASETT 1988). Issues raised included:

- inadequate or missing data
- adverse impacts on existing community and visitors
- the drawcard, to tourists and developers, of the 'undeveloped' environment
- unprofessionalism and inadequacy of the PER
- adverse impacts on the natural environment
- privatisation of public beaches and the bay
- unacceptability of the size and location of substitute beaches
- failure to address alternatives adequately
- doubtful financial viability of the development
- uncertainty as to whether the proposal concept in the PER is the final design
- benefits of marina itself, as opposed to the 'add-on facilities'
- possible design deficiencies.

Under the Commonwealth Environment Protection (Impact of Proposals) Act, PERs do not require the proponent to respond to submissions in a final document.

Attempts were also made to involve the public prior to the PER requirement. In March a public function of 200 local guests was held to discuss the proposal, together with a four-week display, on the island, of the model for the proposal; copies of an audio-visual display; and an insert in a local paper, the *Townsville Bulletin*.

Ministerial assessment and recommendations

On 28 June, following public review of the PER, DASETT requested further information from the proponent. This information was received by DASETT on 1 July 1988. On 12 July recommendations on the proposal, based on an assessment by DASETT, were made by the Minister of the Arts, Sport, the Environment, Tourism and Territories.

These assessment stated that overall 'compliance with the guidelines is sketchy or absent' (DASETT 1988, p. 6) and that information was not always applicable to the current version of the proposals. Moreover, the PER was too lengthy for this level of assessment, contained too many peripheral discussions, and frequently referred to Volume 2 of the IAS which was not released to the public with the PER. It was recommended (DASETT 1988, p. 18) that:

> the Authority not grant a permit for the development of the project within the Great Barrier Reef Marine Park until the proponent has obtained and supplied the Authority with information, further baseline studies and/or modelling on the following:
>
> - coral communities
> - hydrodynamics
> - present suspended sediment communities
> - sediment transport
>   - effects on coral communities
>   - effects on marina and internal beaches
>   - effects on remaining Nelly Bay Beach
> - water pollution
>   - sewerage from marina users
>   - Gustav Creek
>   - runoff and seepage from landscaping
> - basin flushing and resultant water quality effects
> - construction and turbidity effects
>   - from building of breakwaters
>   - of pumping from marina basin/landfill areas
>   - of dredging access channel
>   - of dismantling bund walls
>   - of completing of access to marina basin
> - such other matters affecting the Marine Park as the Authority requires.
>
> and the Authority is satisfied that likely environmental effects disclosed by the information lie within acceptable limits

Other information required by the Ministerial recommendations related to studies into archaeological value, environmental monitoring programs, and information to provide certainty about responsibility for costs relating to rehabilitation and removal of structures in the event of any damage or economic failure (cited in Whitehouse 1992). The recommendations stated that in the event that this additional information was prepared for consideration by the Marine Park Authority, DASETT would consider the object of the Commonwealth Act to be met. The recommendations were publicly advertised on 18 July 1988.

A second PER

Based on the Commonwealth government's assessment, the proponent was required to prepare a second PER on 11 August 1988. Detailed guidelines were prepared by the Marine Park Authority, based on DASETT comments, and the two-volume PER (MKL 1988b) was subsequently submitted by McIntyre and Associates Proprietary Limited (for the proponent) within the same month (MKL 1988). Some of the land-based matters were not addressed, being outside the jurisdiction of the Marine Park Authority, and having been previously considered by the Townsville City Council and the Land Administration Commission.

Public review of the second PER

The PER was placed on public exhibition from 2 to 29 September 1988, with an informal extension to 6 October. A total of 27 submissions were received and summarised by the Marine Park Authority. Many of the issues appear similar to those submitted for the first PER, including:

- issues of use and amenity (for example, lack of a formal attitudinal survey, inadequate information on economic factors, loss of reef flat, increased visitors, changes to lifestyle, and diminished recreational opportunities)
- inadequate facility design and consideration of alternative locations
- insufficient consideration of potential Aboriginal artefacts
- loss of natural environment
- inadequate coral surveys
- water quality (for example, sewage and oil spills)
- inadequate hydrodynamic and sedimentation studies
- inadequate monitoring.

Some submissions also suggested that, rather than restricting assessment to matters under the jurisdiction of the Marine Park Authority, the effects of the whole development should be addressed. The Marine Park Authority also sought independent comments on the PER from specialist reviewers. Many of the comments from these specialist reviewers expressed concern about the adequacy of the PER, and raised similar concerns to the public submissions regarding, among other issues, impacts of water pollution and sedimentation on reef communities, and inadequate hydrodynamic models.

The assessment report

Following the public review period, the Marine Park Authority produced an assessment report in October 1988 (GBRMPA 1988). This report examined only those issues relevant to the Marine Park, and assessed the proposal against the objectives of General Use A Zone (see p. 114). This zoning essentially allowed harbour works and construction of mooring facilities, but only if consistent with the protection of the Great Barrier Reef Marine Park and if permitted by the Marine Park Authority. Within this context, three key areas of concern were identified:

- the extent of potential impacts on the fringing reef in Nelly Bay and adjacent bays
- the impacts on existing and future use and amenity (for example, health and safety aspects)
- the management of a significant archaeological site.

In short, it was concluded that 'the proposal as put is likely to have unacceptable environmental impacts on the GBRMP [Great Barrier Reef Marine Park] and that therefore the Authority could not support the issue of a permit on the basis of the existing proposal' (GBRMPA 1988, p. 21).

The proposal is approved

Following discussions between the Marine Park Authority and the proponent on 26 October, whereby it was agreed to modify the proposal and establish environmental safeguards to limit damage to the environment, the proposal was approved by the Marine Park Authority. The permit was issued on 28 October, subject to an insurance bond to safeguard against project failure. Construction began in June 1989, although this was delayed due to appeals to the Local Government Court and Administrative Appeals Tribunal regarding the approval, all of which were dismissed. The development was discontinued in 1990 following the receivership of the proponent and the subsequent expiry of the permit. However, the insurance, designed to provide a buffer against potential economic failure of the project, was not renewed in time to allow compensation for the damage caused by the half-finished project (Figure 6.4). In this regard, the Marine Park Authority, having required the insurance in the first place, clearly did not monitor its status.

Figure 6.4 Photograph of Magnetic Quay site
Source: photographed by Nick Harvey 1996

An independent review

Faced with the controversial nature of the decision and the extent of residential opposition, the Commonwealth Minister for Arts, Sport, the Environment and Territories (Senator Ros Kelly) triggered an independent review into the decision-making process. Mr John Whitehouse of the law firm Dunhill Madden Butler was selected both to examine the procedures the Marine Park Authority had used to arrive at their decision, and to investigate criticism made by the community group 'Island Voice' and Democrat Leader Senator John Coulter against the Marine Park Authority. The public process involved an invitation for written submissions (to be submitted from 4 February to 28 February 1992 (28 were received), and the organisation of a public round-table meeting on 20 March 1992. However, many of the recommendations arising out of this independent review related more to political issues raised *after* the objects of the EIA process had been fulfilled. Submissions were summarised in the review report (Whitehouse 1992). Those issues different to previous concerns are listed in the box below.

Issues raised in the independent review of the Magnetic Quay proposal

- problems of visual amenity arising from an incomplete site
- impacts on kelp beds
- inadequate assessment of the proponent's financial stability
- the pro-development attitude of the Great Barrier Reef Marine Park and the political nature of the decision that ignored independent scientific information
- problems associated with 'approval in principle'
- unworkable environmental safeguards
- the fact that the zone 'General Use A' was created to enable the marina development to proceed
- lack of public involvement in the intergovernmental working group established in 1986
- the problems associated with the fact that the proposal was allowed to proceed without rehabilitation insurance in place
- breaches of the code of environmental practice
- delays in the EIA process, which led to project failure
- insufficient conditions attached to the permit
- the fact that a permit should not have been issued until solutions were documented and publicly available
- problems associated with requiring developers, rather than impartial bodies, to undertake EISs

During the independent review the Queensland Tourist and Travel Corporation, the Advance Magnetic Island Association (330 residential members), and the Townsville City Council all voiced support for the proposal and for the decision of the Marine Park Authority. The Townsville City Council suggested that 'no environmental detriment would result from the project' (cited in Whitehouse 1992, p. 6). Issues arising from the round-table meeting were similar to issues raised in the assessment process. Additional issues included the low bond (guarantee to cover the liability of the proponent) required of the proponent and the need to establish mechanisms to coordinate the EIA process where numerous jurisdictions are involved.

The report of the independent review concluded that so long as 'the proposal is properly justified, subject to adequate environmental assessment and that development works are properly supervised and their impact properly monitored', the marina was not an inappropriate development within the zoning (Whitehouse 1992, p. 93). While the report of the independent review gave support to the decision, it also emphasised the need for adequate insurance.

The report also suggested that while it was unreasonable to expect the Marine Park Authority to assess the economic viability of the proposal or the financial status of a proponent, it was necessary to assess the implications in the event of financial difficulties and to utilise bonds, bank guarantees, and insurance as guarantees. It was also noted that, given its circumscribed jurisdiction, the Marine Park Authority was restricted in its ability to assess the impacts in the Marine Park. The report also recognised that

> . . . the fragmented nature of the consideration of the project involving as it did the overlapping responsibility of at a minimum one Commonwealth agency, two State agencies and the Local Municipal Authority raised many legitimate concerns as to whether all of these aspects have been considered and examined in relation to the entire project. Moreover the fact that such important matters concerning the future planning and environmental management of Magnetic Island required resolution in the context of a specific development proposal, rather than being the subject of a forward regional planning study unrelated to a specific development proposal, gave rise to additional concerns.
>
> Whitehouse 1992, p. 56

Other suggestions in the report included the development of guidelines for environmental surveys of marine environments. The purpose of these guidelines would be to overcome the potential for ambiguity in such surveys, thereby assisting environmental assessment procedures. Also suggested was the need to avoid giving 'approvals in principle', as they could create ambiguity and could lead to misinterpretation.

A new proposal

As the site was only partially developed, in 1993 the Queensland Department of Lands (previously the Lands Administration Commission) sought expressions of interest for development on the Magnetic Quay site. In order to expedite the process, the department lodged an application for a permit with the Marine Park Authority that, if successful, would be transferred to the successful developer (this made the department the effective proponent until such time as the permit could be transferred to the future developer). Five applications were assessed, resulting in the selection of Nelly Bay Harbour Proprietary Limited on 10 August 1994. An EIS and an IAS was required under both Commonwealth and Queensland EIA legislation respectively.

The 1995 proposal (Nelly Bay Harbour) was different again in emphasis, with plans for a residential waterfront development with 'low key tourist and commercial facilities' relative to the previous proposals. Key components of the development include:

- a ferry terminal
- a barge terminal
- a commercial complex (including retail)
- a bus interchange
- a taxi rank
- a tavern, motel
- a boat ramp
- marina berths (30-day use)
- residential development (88 lots)
- public open space
- public waterways
- a sedimentation basin (at Gustav Creek)
- a sewerage system
- pump-out facilities for boats
- a stormwater drainage system.

Development would also entail construction of breakwaters, reclamation, and a dredged access-channel.

It is interesting to note that the previous Commonwealth requirement was for a PER level of assessment. In this case, however, an EIS was considered more appropriate because the development was different and new issues were evident. For example, the proposal falls within a World Heritage Area and development needs to be in sympathy with such an area. Although this was also the case in 1986, it was now a more important issue to the public. Moreover, the proposal aims to construct more permanent residential development in the Great Barrier Marine Park—the previous proposal had not had permanent residential accommodation (CEPA, pers. comm., 1995).

In apparent recognition of the problems of duplication (assessment could, conceivably, have been required at both the Commonwealth and the Queensland level), it was decided to undertake a joint assessment process.

A direction for an EIS was made on 22 December 1994 (Sinclair Knight Merz 1995). In March 1995 an advisory body was set up with the aim of reviewing the progress of the assessment (CEPA, pers. comm., 1995). However, the Queensland Department of Premier, Economic and Trade Development was responsible for the day-to-day running of the EIA process. Terms of reference for the EIA were produced in February 1995, and although no formal advertisements were made for public comment, consultation was sought with relevant state and Commonwealth authorities, local government, conservation councils, and major community organisations (for example, Island Voice) over a 4–5 week period (CEPA pers. comm., 1995). In September 1995 Sinclair Knight Merz published the EIS on behalf of the proponent, the Department of Lands (Sinclair Knight Merz 1995).

It was decided that the 28-day public review period under the Commonwealth legislation was inadequate for this type of proposal, and the EIS was released for a six-week public review period from 11 October to 22 November 1995 (Sinclair Knight Merz 1995; Commonwealth Environment Protection Authority, pers. comm., 1995) (note that public review is discretionary under the Queensland legislation). In addition, a public meeting was held during this period (CEPA, pers. comm., 1995). A total of 250 submissions were received (80–90 were 'form' letters), the majority of which were opposed to the proposal (CEPA, pers. comm., 1995). Essentially, opposition was not aimed at the development of the site but rather at the type of development: it was considered too commercially oriented, too high in density, and inconsistent with existing development on Magnetic Island. The community organisation Island Voice was totally opposed to the proposal. At the time of writing the proponent was responding to the submissions in a supplement.

SOUTH AUSTRALIA

There are detailed figures on the practice of EIA under the South Australian Planning Act, from 1982 to 1994 (Harvey 1995). There have only been three EISs (plus an amendment to an earlier EIS) completed under the Development Act, which came into force in 1994, and one of these was a joint assessment under Commonwealth legislation. Between 1982 and 1994, under sections 49, 50, and 59 of the Planning Act, the Minister for Environment and Planning called for a total of 60 EISs to be prepared. Just under half (27) of these have been for marina proposals. Of the total number of EISs, five were required for tourist-related projects; five for industrial projects; four for power-generation-related projects; four for electricity-transmission-line projects; four for transport projects; three for mining projects; three for water management projects; two for waste-management projects; two for urban development projects; and one for a defence project (Harvey 1995). By 1997 there were 41 EISs completed for various projects. However, a recent 1996 amendment to the Development Act introduced the PER and the development report (DR) as two new levels of assessment in the South Australian EIA process. Given the low number of EIS examples under the Development Act, the EIS case study provided here has been selected from proposals under the Planning Act. It is a development that has attracted ongoing national attention. Much of the material on this case study has been drawn from Harvey (1996).

The Hindmarsh Island bridge and marine extensions EIS

The proposal

Hindmarsh Island has recently received considerable media attention because of the Aboriginal heritage issues involved in the proposed developments, specifically the 'secret women's business' that could be affected by a proposed bridge across the River Murray. However, the bridge was only part of a larger marina and residential development project that was subject to EIS requirements. The private marina development proposal (by Binalong Proprietary Limited) became quite complex because it was actually an extension of an existing marina development, in relation to which there was some confusion over previous approvals. The main elements of the proposal were:

- a bridge to Hindmarsh Island
- a 1150-berth marina (about half of which had already been approved and one-quarter constructed)
- 500 dry-stand berths (including 200 already approved)
- 859 residential allotments (including 324 waterfront allotments)
- associated waterways and lagoons
- a tourist centre
- a sewage treatment plant
- a water-filtration plant (see Figures 6.5 and 6.6).

The EIS and public review

Although the marina extensions in themselves were not considered a major impact, the associated residential development and proposed bridge to Hindmarsh Island gave rise, on 24 October 1989, under the Planning Act, to a requirement for an EIS. Because the proponent's consultants (PPK Consultants) had already commenced impact studies, the draft EIS had the shortest proponent preparation time for a draft EIS in South Australia (Binalong 1989).

Issues raised in the Hindmarsh Island EIS

- impact on the Chrystal Street bridge
- traffic approach
- water quality/pollution
- bridge aesthetics
- bridge financing
- size of development
- demand for the development
- alternative bridge locations and alignment
- impact on bird-life and conservation areas
- impact on the rural character of the island
- impact on heritage
- effluent/stormwater treatment/runoff
- impact on local government and rate payers
- visibility of development
- conflict with the state Development Plan

Public comment on the draft EIS was invited through advertisements in state and local newspapers, and the draft EIS was available for viewing at six locations and for purchase at two. A public meeting was held on 4 December 1989 to discuss the EIS process. The document was on display until 18 December 1989, during which time there were 73 public submissions.

Just over half of the 73 public submissions were in favour of the bridge component of the project, although there were local concerns about its potential impact on traffic in the town of Goolwa, its aesthetic impact, and its financing. In addition there was concern about the size of the proposed residential development and about the potential for the bridge and the residential development to change the rural character of the island. There was no clearly identified submission representing Aboriginal heritage concerns. Government comments related to:

- the demand for allotments

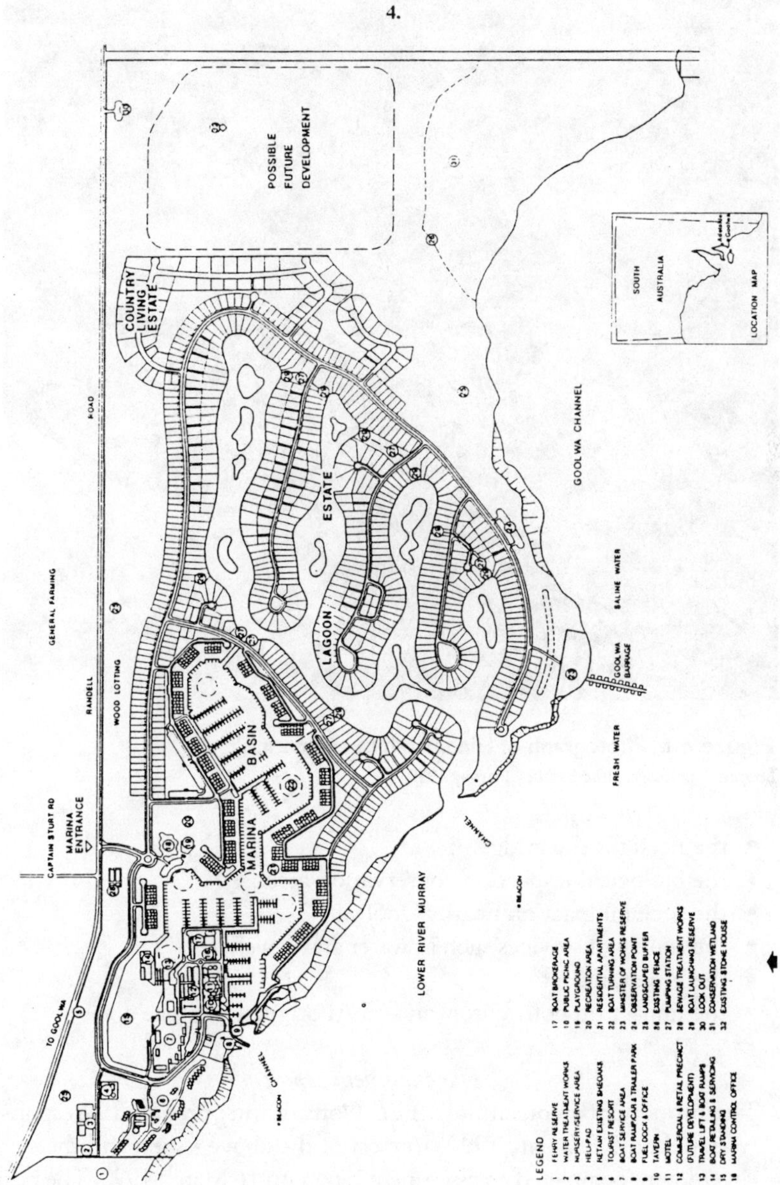

Figure 6.5 Hindmarsh Island proposal

Source: SADEP 1990, p. 4.

Figure 6.6 Photograph of Hindmarsh Island marina
Source: photographed by Nick Harvey 1990.

- the need for the bridge
- the biological impact on conservation areas
- the social impact on nearby Goolwa
- the supply of services such as water and sewerage
- the visual impact
- the impact on both European and Aboriginal heritage areas.

The assessment report

The proponent's consultants, PPK Consultants, prepared a response (released on 5 February 1990) to most of the above issues, and the state government completed an assessment report on 16 March 1990. The principal concern of the assessment report related to the bridge alignments, financial arrangements, the visual impact, and the impact of improved access to the island (SADEP 1990). It should be pointed out that the assessment report contained recommendations regarding the need for appropriate consultation with the relevant Aboriginal representative bodies (recommendation 2.5, SADEP 1990, p. 40). However, the issue of whether or not a bridge was needed had essentially been removed from the assessment process because the South Australian Cabinet had already approved, in principle, the concept of a bridge and the contribution of funds towards it. There were also concerns about the long-term viability of private water and sewerage schemes, and, more importantly, about the way in which the whole project was to be conducted in stages.

The Governor's decision

The Governor assumed control over the development, becoming the decision-making authority, under section 50 of the Planning Act. Following the section-50 declaration over the development area (including the bridge and its approaches), the project was approved by the Governor, subject to certain conditions, on 12 April 1990. These conditions were:

- to conduct appropriate consultation with Aboriginal groups
- to stage the development (with only stage 1 permitted before bridge construction)
- to supply reticulated water.

Another safeguard incorporated into the Governor's approval was the vesting of management and maintenance responsibilities with the local council. The EIS process and subsequent planning approvals for both Goolwa marina and the Hindmarsh bridge project were processed in record time.

Public review of the supplementary development plan

As noted by Harvey (1996), apart from the opportunity for formal public input on the EIS, there was further opportunity for formal public submissions on the proposed supplementary development plan for Hindmarsh Island. This supplementary development plan principally contained policies to establish a marina zone, thereby enabling Binalong Proprietary Limited to proceed beyond stage 1 of its development. The supplementary development plan was placed on public exhibition from 18 October until 18 December 1990. A total of 31 public submissions were received, ten of which came from government agencies. Over half of the submissions (17) were in favour of the supplementary development plan (with minor alterations), and six of these submissions highlighted major concerns and called for substantial changes to the plan. The main issues raised were that rezoning would force shack owners to relocate to the 'holiday house' zone, and that, outside the proposed residential marina site, the supplementary development plan was too restrictive, tending to favour one developer (Binalong Proprietary Limited) over all others. Aboriginal concerns were not raised as a major issue in responses to the planning framework, contained in the supplementary development plan. The supplementary development plan was approved for authorisation by the Minister for Housing and Urban Development on 21 July 1993, subject to execution of the contracts for the Hindmarsh Island bridge. In July 1993 the Minister's advisory committee on planning recommended that, should the bridge contract not proceed, another supplementary development plan should be prepared.

Developments since the Governor's approval

Following the formal public submissions on the draft EIS and the approval, by the Governor, of the Hindmarsh Island bridge and marina in April 1990, there has been significant public opposition to the project, and a number of government inquiries at various levels have occurred. In March 1991 the state government agreed to back the project using savings from the ferry (which would no longer be needed once the bridge was built). The government went further and agreed to fund and build the bridge with a contribution of $3 million (made up of ferry savings), the remainder to be put up by Binalong Proprietary Limited after it had repaid its other loans.

Stage 1 of the development, which was not conditional on the bridge completion, was finished in October 1991. In a clear demonstration of political support, stage 1 was opened by the then Premier of South Australia, John Bannon. However, Stage 2 was conditional on the bridge construction, against which there was mounting public opposition. On 26 November 1991 a petition opposing construction of the bridge was signed by 5043 South Australians and submitted to the government. In light of the opposition to the bridge, the South Australian Department of Road Transport commissioned a feasibility study into options for improving traffic access to Hindmarsh Island. The findings of this report, published in June 1992, recommended a high-level bridge, as opposed to other transport options such as ferries or a tunnel.

In 1993 both an anti-bridge lobby group and a pro-bridge lobby group were formed. In response to growing public opposition to the bridge, the matter was referred to the Parliamentary Environment, Resources and Development Committee, which recommended on 9 September 1993 against a bridge and supported an upgraded ferry service. On 5 October 1993, notwithstanding public opposition to the project and the recommendations of the committee, the government signed a contract with Built Environs Proprietary Limited to build the bridge. In October 1993 the local Goolwa Ratepayers Association, and other groups, expressed support for the findings of the Parliamentary Environment, Resources and Development Committee Inquiry and encouraged the local community, the Construction, Forestry, Mining and Energy Union (CFMEU), and the Building Trades Federation to cooperate in opposing construction of the bridge.

A coalition of various groups opposing the bridge became known as The Kumarangk Coalition. This coalition coordinated a campaign, in the name of the Ngarrindjeri nation, against the bridge development. On 28 and 29 October 1993 a number of community groups established a community picket. Almost all of those involved were local people from Goolwa and Hindmarsh Island. Other groups included the CFMEU; the Conservation Council of South Australia (CCSA); the Lower Murray Aboriginal Heritage Committee (LMAHC); and Greenpeace. The community picketers acted to prevent the commencement of preliminary works for the building of the bridge. The picket was organised to protect Amelia Park, a significant Aboriginal site, which had been identified as archeologically important in an earlier survey (Edmonds 1989). Subsequently, the Minister for Transport provided $20 000 to fund a comprehensive survey, under the direction of the Minister for State Aboriginal Affairs, of the bridge site and the surrounding area. By 11 November all work on the bridge had stopped because of industrial action. On 9 December a petition of 2000 signatures opposing the bridge was presented to the government.

The bridge issue remained unresolved in the lead-up to the state elections in December 1993, when there was a change of government. The new Liberal government had opposed the bridge when in opposition, referring to it as the 'bridge to nowhere'. Once in government the Liberals immediately suspended work on the bridge and set up an inquiry headed by Queen's Counsel Justice Jacobs to examine the financial and legal implications of the bridge contracts. Despite requests by groups such as the CFMEU, the CCSA, the Aboriginal Legal Rights Movement (ALRM) and the LMAHC to expand the terms of reference to include Aboriginal heritage issues and environmental issues, the Jacobs report remained focused on the contractual arrangements.

In February 1994 the Jacobs inquiry confirmed that the government was contractually obligated to build a bridge and that, in the event that they failed to do so, they would face compensation costs of $12.5 million. The report, which was not publicly released, also recommended investigation of alternative sites. The government subsequently ruled out any alternative sites and, on 15 March, lifted the stay on the bridge construction. Further public protests staged at the bridge site frustrated the developers Binalong Proprietary Limited. On 29 March 1994 Binalong obtained an interim injunction, under section 45D of the Commonwealth *Trade Practices Act 1975*, from the Federal Court of Australia. Binalong also sent letters to a number of bridge protesters, threatening to sue them for the company's increasing debt and loss of profit.

On 3 May, the State Minister for Aboriginal Affairs, having received a detailed archaeological survey prepared for the government, gave conditional authorisation (under section 23 of the *Aboriginal Heritage Act 1988* (SA)) to disturb Aboriginal sites. Work recommenced on the bridge but encountered public opposition resulting in a number of arrests.

The next stage in public involvement was the turning point in the opposition to the development project. On 9 and 10 May 1994 (over four years after the EIS process had been completed and project approvals granted), Aboriginal women met to discuss the significance of the area to Ngarrindjeri women. Two days later, on 12 May, the then federal Minister for Aboriginal and Torres Strait Islander Affairs, Senator Robert Tickner, made an emergency declaration under section 9 of the *Aboriginal and Torres Strait Islander Heritage Protection Act 1984* (Commonwealth) that effectively halted any bridge construction for a period of 30 days, thereby allowing for a detailed report to be prepared. On 12 May 1994 the Minister appointed Professor Cheryl Saunders to prepare a report under section 10 of the Aboriginal and Torres Strait Islander Heritage Protection Act. Professor Saunders was required to examine the 'significance of the area to Aboriginals and the nature and extent of the threat of injury to, or desecration of, the area and the effect the making of a declaration may have on other persons' (Saunders, quoted in Harvey 1996, p. 45).

In the same month the issue of civil liberties was raised regarding the injunctions placed on various protesters and community organisations. On 13 May a detailed letter of complaint was sent to the United Nations Human Rights Committee in Geneva, claiming violation of the *International Covenant on Civil and Political Rights*. It was claimed that the interlocutory injunctions placed on seven respondents (CFMEU, the organisation Friends of Goolwa and Kumarangk, and five individuals) was a violation of Article 19(2) of the Covenant.

Public opposition intensified, and on 5 June approximately 800 Aboriginal people from across the state met near the proposed bridge site.

It became clear that the complexity of the issue was such that more time was needed for the section 10 report being prepared by Professor Saunders. The Minister extended the section 9 declaration (this section enabled the section 10 report) for another 30 days.

The Saunders report was completed on 7 July 1994 and was presented to Senator Tickner. This report suggests that Ngarrindjeri women were not aware of any discussions about the development proposal until late March 1994 (Saunders 1994, p. 31). Of particular importance was the attachment, to the report, of two envelopes containing what has been called 'secret women's business' outlining the significance of the area to Aboriginal women. The contents of these envelopes, marked 'confidential' and 'not to be read by men', were critical in the Minister's decision to place a 25-year ban, taking effect from July 1995, on the building of the bridge. In a media release on the same day the Minister referred to the injunctions under the Commonwealth Trade Practices Act, and expressed a wish that these could be lifted. They were lifted some time later, on 14 October, thereby removing an obstacle in the path of public involvement in the debate.

The meeting on 5 June did not end public involvement in the development proposal. On 15 February the Federal Court, in a decision on an action brought against the state government by Binalong Proprietary Limited, overturned the Minister's bridge ban. The national debate over the bridge issue intensified in March 1995 when the then Shadow Minister for the Environment, Brian McLaughlin, tabled, in Federal Parliament, a number of the confidential documents from the Saunders Report. He claimed that a breach of security in the government's handling of the report had led to men seeing the information that was supposed to be seen only by women. Political pressure and public outrage forced him to resign as Shadow Minister on 10 March 1995.

In May 1995 the whole issue of the bridge ban was thrown into confusion when a breakaway group of Ngarrindjeri women denounced the authenticity of the 'secret women's business'. This led the South Australian government, in June 1995, to announce a Royal Commission into the affair. At the same time the Federal Minister for Aboriginal and Torres Strait Islander Affairs, Robert Tickner, announced a separate but broader inquiry—to follow the Federal Court appeal decision (which it subsequently lost). In the meantime, the South Australian government's Royal Commission has encountered significant public opposition from, among other groups, church groups questioning the validity of a commission of inquiry into people's beliefs. Notwithstanding this opposition, the Royal

Commission concluded that there had been fabrication of evidence by the Ngarrindjeri women. Most recently, in May 1997, the Commonwealth government passed legislation to exempt the bridge development from the Commonwealth Aboriginal Heritage Act, thus avoiding the need for another inquiry. This legislation paves the way—after four inquiries—for the construction of the bridge. However, a legal challenge against this legislation has been initiated by Aboriginal groups and, together with future compensation claims from the developers, the Hindmarsh Island EIS saga is likely to continue into the next century.

An important point that comes out of the Hindmarsh Island case study is the fact that, although there was opportunity for public involvement through the EIA process and public comment on the associated amendments to the development plan, issues concerning Aboriginal heritage were not raised through either of these processes. This suggests that greater care, and perhaps alternative EIA public consultation methods, are necessary for dealing with such matters as 'secret women's business'. However, the key problem is to unearth such issues at the appropriate stage—not years after approvals are given, as happened in the case of Hindmarsh Island.

TASMANIA

Current practice in Tasmania is difficult to determine given the recent changes to the process and overall planning system. Practice prior to the new Environmental Management and Pollution Control Act has been summarised by Thomas (1995) in terms of numbers of EISs. However, given that the Department of Environment and Land Management does not, as yet, have a centralised database of EIAs (and are currently in the process of establishing these records), the value of these figures is unclear (P. Sheldon-Collins, pers. comm., 1996).

The Mt Lyell mining redevelopment environmental management plan (EMP)

The proposal

In December 1994, following a one-hundred-year history of mining at Queenstown, mining operations ceased with the closure of the Mount Lyell Mining and Railway Company (CMT 1995). Despite the fact that significant environmental impacts had resulted from these operations, it also became clear that the closure would result in serious economic and social impacts. Surrounding townships had been established around mining operations and the closure meant population decreases, unemployment problems and significant economic losses (CMT 1995).

There was, however, still potential for redevelopment of the mine, and in this context the Tasmanian government, as the owner of the resource, sought expressions of interest from companies to continue mining operations. The government subsequently established an agreement with Copper Mines of Tasmania (a subsidiary of Gold Mines of Australia) on 28 June 1994. This was later formally signed on 13 October and was ratified by the *Copper Mines of Tasmania Pty Ltd (Agreement) Act* on 16 December 1994. As part of this agreement Copper Mines of Tasmania had committed $5 million to undertake an economic, engineering, and environmental study by April 1995. Copper Mines of Tasmania acquired the site in late December 1994 and was granted a mining lease by the government. They were also granted an indemnity against impacts from past operations, identified in a previous preliminary environmental review of the site in November 1994.

Copper Mines of Tasmania planned to increase mining of the ore deposits from the amount of previous operations (1.7 million tons per year) to between 2 million and 3.5 million tons per year. This was to be done via a 'sub-level caving' method. The mine (Figure 6.8) was predicted to have a life of 15 years, although continued exploration of the region was also proposed. In order to proceed at this rate, existing infrastructure needed to be upgraded (for example, underground crushers, mine ventilation, and decline extensions). In addition, two grinding mills, a slurry pipeline, an open ore-stockpile, and a major tailings dam were planned. Among other facilities, new access roads, waste-rock storage, management facilities, and product handling and shipping infrastructure were proposed. The mine was originally expected to operate 24 hours per day, seven days a week. Expected costs were $900 million, and the mine was expected to produce 400 000 tons of copper and 500 000 ounces of gold in the first ten years. In March 1995 relevant licences under the Environmental Protection Act were transferred from the previous operator to Copper Mines of Tasmania, as required by the Copper Mines of Tasmania (Agreement) Act.

Triggering

Pursuant to the Copper Mines of Tasmania (Agreement) Act, and in accordance with section 18 of the State Policies and Projects Act, the proposal was declared to be a project of state significance. The licence approval process would normally have been the responsibility of the

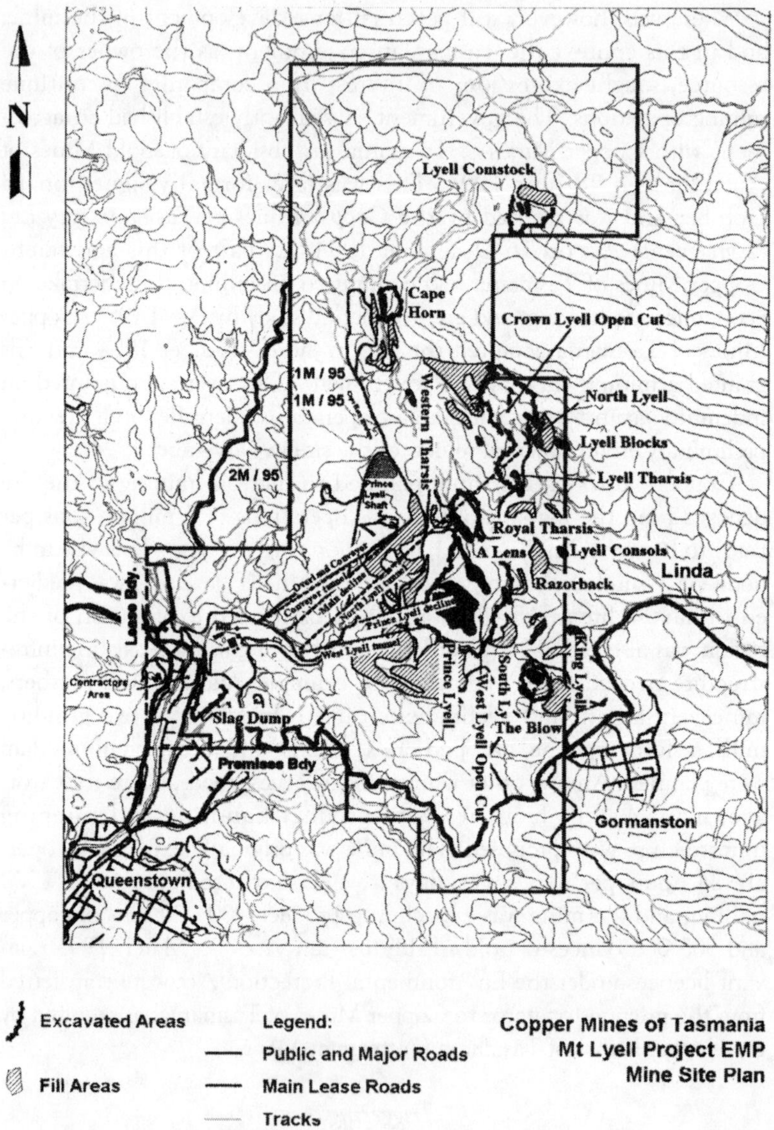

Figure 6.7 Mt Lyell mining proposal site plan
Source: CMT 1995

Figure 6.8 Photograph of Mt Lyell mining site
Source: supplied by Copper Mines of Tasmania Pty Ltd

Director of Environmental Control, Department of Environment and Land Management, but because the mining operations were a scheduled activity and therefore required a licence under the then Environment Protection Act, this responsibility was transferred to the Sustainable Development Advisory Council. On 6 February 1995, in accordance with the provisions of the State Policies and Projects Act, the Tasmanian Minister for Environment and Land Management required an integrated-assessment process. This direction was subsequently approved by both houses of the Tasmanian Parliament on 7 February.

However, the integrated-assessment process was slightly modified by the Copper Mines of Tasmania (Agreement) Act. Most notably section 7(3) of the Act precludes the SPPA from 'preventing CMT from undertaking the CMT Proposal before an order [decision by the Governor] . . . is made'. Under normal circumstances the proposal could not proceed until the assessment was complete and a decision made (section 19, State Policies and Projects Act). An amendment to the Copper Mines of Tasmania (Agreement) Act in 1995 also enabled 'preparatory works' to proceed, such as an initial tailings dam at Princes Creek, exploration drilling, decline extensions, and ore and waste-rock storage. The Tasmanian Sustainable Development Advisory Council noted that

> . . . the establishment of the Tailings Dam . . . was to enable a facility to be in place and ready to accept and treat tailings immediately upon the approval being granted for the Project to proceed. In the event that the CMT Project did not proceed, or was refused, the dam would be dismantled and the land rehabilitated . . .
>
> SDAC August 1995a, p. 15

It is possible that such preparatory works, and the associated costs invested by August 1995 ($22 million), constituted indirect pressure on assessment- and decision-making authorities to approve the project, particularly as it was the government that initiated tenders for the continuation of mining in the first instance. Initially, a licence application and an EIS had been completed in the form of a Development Proposal and Environmental Management Plan (under the old EIA procedures), but these were withdrawn in light of the *Copper Mines of Tasmania (Amendment) Act 1995* (CMT 1995).

Scoping

Nevertheless, as part of the Ministerial direction to the Sustainable Development Advisory Council, the broad scope of the assessment was outlined pursuant to section 20 of State Policies and Projects Act—section 20 requires the Minister to outline the 'matters to be addressed . . . the process to be followed, and the time within which the . . . assessment must be completed'. A completion date for the process, 31 August 1995, was set by the Minister, resulting in a six-month framework. This appears to be a tight time-frame relative to practice in other jurisdictions, particularly as the project is of state significance. For example, most EISs in South Australia extend over 50 weeks (Harvey 1995), and the other case studies summarised here involved processes extending to at least 12 months (for example, the Sydney Harbour Tunnel, the third Runway, and Magnetic Quay).

Project-specific guidelines were also produced, by the Sustainable Development Advisory Council, for Copper Mines of Tasmania, and these were released for public comment on 11 February 1995. This was a significant practice in that public review of the scoping procedure is not formally required by the State Policies and Projects Act. These guidelines were subsequently amended and forwarded to Copper Mines of Tasmania on 8 March 1995.

The environmental management plan (EMP)

Rather than an EIS as such, the proponent was required to produce its equivalent in the form of an environmental management plan (EMP) that aimed to provide 'a framework for Best Practice Environmental Management' (CMT 1995). This document is unique in that many EISs in Australia are static documents, being used only once in the decision-making process, with the possible exception of Western Australia's environmental review and management plans (ERMPs). In contrast, it was proposed to subject the EMP to continuous review and monitoring, even after the final assessment and determination.

The first draft of the EMP was completed in April 1995 by Thompson and Brett Consulting Engineers (Thompson & Brett Consulting Engineers 1995), approximately two months after the Ministerial direction, and one month after receipt of the Council's guidelines. Key contents (of the revised version) include:

- the project rationale
- a site history
- a description of the existing environment and the proposed mining development
- an outline of the development and production processes
- an environmental management plan addressing
  - issues of flora and fauna
  - archaeological issues
  - cultural and heritage issues
  - geological and soil impacts
  - recreational issues
  - water and air pollution
  - erosion problems
  - risk management
  - traffic and aesthetic impacts
  - waste management strategies
  - monitoring and review programs.

Alternatives were also considered, to a limited extent, in accordance with the Sustainable Development Advisory Council's guidelines in relation to mining techniques and waste-management options. There did not, however, appear to be any explicit consideration of the 'no go' option. This omission is probably a reflection of the type of project: mining operations have little choice but to follow the location of mineral deposits. Moreover, the project appeared to be strongly justified in light of the significant economic benefits to the west coast of Tasmania and to Tasmania as a whole, as well as the environmental benefits resulting from improved operations and the reduction of impacts from past mining activities.

Public review of the EMP

In April 1995, as required by the Copper Mines of Tasmania (Agreement) Act, the Sustainable Development Advisory Council reviewed the adequacy of the EMP. This review resulted in minor revisions to the document within the same month. The EMP document was then released for a public-review period of 28 days. Given that public review for this component of the EIA process is not specified in the State Policies and Projects Act (even the requirement for an EIS or EMP (or equivalent) is not explicit in the Act), this was a significant undertaking.

Nineteen submissions were received, only five of which were from the public—14 were from agencies (for example, government departments, the Australian Nature Conservation Agency). All submissions were summarised by the Sustainable Development Advisory Council. Major issues raised in this summary and issues raised during public review of the draft assessment report are outlined below. Most public concerns related to the tailings dam and water management.

The draft assessment report

Following a review of the EMP, the Sustainable Development Advisory Council produced a major issues paper outlining topics omitted in the EMP. On 2 June 1995, following a meeting between the Sustainable Development Advisory Council and Copper Mines of Tasmania, the latter submitted a formal response to the major issues paper. Then, in June 1995, the Sustainable Development Advisory Council prepared a draft integrated-assessment report (SDAC 1995a), as required by the State Policies and Projects Act. This was also released for a public-review period of 28 days, commencing in June and ending on 24 July. Such a legislative provision for review at this stage of the process is unique in Australia, as those jurisdictions with provision for government assessment reports or equivalent do not have similar provisions for public comment.

However, a relatively small number of submissions were received—only 16. Moreover, the proportion of these that were comments from the public was again small (4). This opportunity for public comment was supplemented by four public hearings, held by the Sustainable Development Advisory Council in late July and early August. These hearings resulted in 22 oral submissions. Key issues raised in these submissions include comments on the EMP and the draft assessment report. Again, the major concerns related to the tailings dam and to water quality management.

Key issues raised in submissions responding to the Mt Lyell Assessment Report

- impacts from the tailings dam—for example, from its design and siting
- water quality issues
- inadequate hazard/risk analysis
- tourism growth-opportunities as well as conflicts with tourism
- heritage preservation
- revegetation issues
- health issues
- benefits to, and impacts on, the community
- impacts on transport
- unclear baseline situation
- waste-rock dumps and increased pollution resulting from these
- issues relating to flora and fauna

Revision of the EMP

Following the review period, the Sustainable Development Advisory Council asked Copper Mines of Tasmania to revise the EMP in light of both the written submissions on the draft integrated-assessment report and the oral submissions resulting from the public hearings. Copper Mines of Tasmania submitted a revised EMP on 11 August 1995. The Sustainable Development Advisory Council initiated an independent review of the revised document using consultants AGC Woodward Clyde. This review concluded that the EMP was a 'competent and comprehensive document' (SDAC 1995a) requiring only minor changes. Subsequently, an addendum to the EMP was required by the Council and produced by Copper Mines of Tasmania on the 23 August. The EMP was finally approved on 25 August 1995, only six days before the date initially set for the final assessment completion.

The final assessment report

The Sustainable Development Advisory Council produced a final integrated-assessment report (SDAC 1995b) based on: all assessment documents; public and agency submissions; and advice from the Department of Environmental and Land Management and various independent consultants. The project was also assessed in the context of the new South Australian Resource Management and Planning System objectives of sustainability, and the national strategy on ecologically sustainable development (see p. 193). Overall, the Sustainable Development Advisory Council concluded that these objectives had been 'properly pursued' in the assessment, and that the project should proceed. In addition, 25 recommendations (or 'Council positions') were outlined. These covered issues and licence conditions relating to a wide range of matters, including:

- maximisation of employment benefits
- allocation of royalties to environmental remediation
- coordination with tourism ventures in the region
- re-vegetation
- mechanisms for heritage management
- ensuring minimal impacts during construction
- adherence to noise and vibration levels
- tailings-dam construction conditions
- mine-water management
- waste management.

The decision

The Council's recommendations were forwarded to the Minister on 31 August 1995, thereby meeting the overall time-frame for assessment. In early October the Governor of Tasmania made an order to approve the project (Jones, pers. comm., 1996), and there were no appeal rights on this decision (Jones, pers. comm., 1996).

Comments

Although the assessment process was undertaken within a very tight time-frame, it was a notable example of EIA in Australia, particularly in light of the provisions for public involvement at three stages of the whole assessment process (at scoping, review of the EMP, and review of the government's assessment). Moreover, the practice of assessing the adequacy of the EMP (as formally occurs in Western Australia and Victoria) is an important check in the process, as is the emphasis on monitoring and continuous review of the EMP, with the latter ensuring that the EMP does not become a static document. These factors appeared to strengthen the monitoring and review components of EIA, which have often been described as weak, particularly in terms of monitoring (Buckley 1989). However, one may also question whether or not it was a foregone conclusion that the proposal would proceed in light of its apparent benefits. The following factors, among others, pointed towards a favourable outcome for the proponents

- Mining operations had already been undertaken in the location for one hundred years.
- Surrounding communities had been based on these mining operations, and hence would be adversely affected by its closure and were unlikely to object to the proposal.
- The environmental impacts from past operations were expected to be reduced, resulting in a general improvement to the surrounding environments.
- The financial benefits would include significant contributions to the local and Tasmanian economy.
- The government was the resource owner, and initially sought expressions of interest from mining companies to redevelop the mine.

Moreover, preparatory works amounting to an investment of $22 million had been approved and commenced *prior* to the final decision. In this context, then, the decision-making process may have been concerned more with mitigating and preventing environmental impacts, rather than considering the 'no go' option.

VICTORIA

Given the discretionary nature of the Victorian Environment Effects Act, limited numbers of assessments are undertaken each year (approximately ten) (Kinhill 1994; Environment Assessment Branch 1995). Since 1975, 153 assessments of EESs have been completed, while 17 EESs have been produced without assessment (Environment Assessment Branch, pers. comm., December 1995). Prior to 1990 the majority of assessments were for public works (53 per cent), although increasing numbers of private works are being referred for assessment (Environment Assessment Branch 1991).

Case study of the Mornington marina EES

While there has been a long-acknowledged need for recreational boating facilities and safe boat havens in Port Phillip Bay, development has been slow due to community concern about development proposals, high infrastructure costs, and dissatisfaction with decision-making processes. In this context, in August 1997 the Victorian government established a new process for assessing coastal proposals to account for environmental and community issues. The three major features of this process are:

- establishment of an inter-departmental coastal task group to identify appropriate projects and sites for coastal developments
- an investigation, by the task group, into four potential marina sites in Port Phillip Bay
- the establishment of a two-stage process to streamline approvals.

Stage one of this new approval process involved a number of steps, including the establishment of a local working-group for each location; an early period of community consultation to identify the relevant issues for the study; a study of the environmental, social, and economic implications of development (the EES); the public exhibition of the study; and assessment of submissions to an independent panel. Due to concerns about harbour safety, triggered by severe storms in the mid 1980s, Mornington Harbour was selected as a potential development site. The Mornington Working Group was established in 1987 and included representation from the Shire of Mornington; the Ministry for Planning and Environment; the Port of Melbourne Authority; and the Department of Conservation, Forests and Lands.

The preferred development concept

Based on a comparison of alternative designs, a 'preferred development concept' (see Figure 6.9, p. 147) for the harbour was proposed entailing:

- extending the existing jetty by 14 m
- adding a right-angle arm (150 m) to the existing jetty (instead of a rock breakwater)
- provision of mooring facilities for 150 boats (floating pontoons)
- facilities for pumping out boats
- a restaurant
- increased traffic access
- parking facilities.

It should be noted, however, that this was conceptual only: more detailed assessment would be required in stage 2, following the selection of a developer and the production of a specific proposal fitting the constraints identified in stage 1. An independent panel inquiry gave full support to this staged approach to assessment, noting that 'it allow[ed] a full assessment of the ideas put forward without commercial pressures dominating the discussion . . . it genuinely revolved around safety and environmental issues' (Panel Report 1991, p. 10). The panel's report also noted that the staged approach provided greater certainty to the developer who would be chosen in stage 2.

Requirement for an EES

Following consultation with the Minister for Environment and Planning, the Mornington study was required to be undertaken in the form of an EES, following the provisions of the Environment Effects Act. The EES was commissioned jointly by the Shire of Mornington and the state government (specifically, the Department of Planning and Urban Growth and the Department of Conservation and Environment). As often occurs in the Victorian EIA process, the EES was undertaken simultaneously with the preparation and exhibition of an amendment to the Mornington Planning Scheme. This scheme, which was prepared by the Shire of Mornington, created the Mornington Harbour Zone under the Planning and Environment Act 1987. The EES was intended to be read jointly with the scheme. The final decision by the Shire on the amendment was dependent on the outcome of the EES. If the amendment were to be approved, any subsequent harbour development would require a town-planning permit.

Public review during EES preparation

In terms of the EES, a study brief for consultants Loder & Bayly was prepared by the Shire of Mornington and the state government. Public involvement was identified as a condition of any identification, in the EES, of relevant issues. Thus, although not legislatively required by the EIS process (in practice, 28 days are usually set for public scoping), public consultation was initiated early by the consultant as part of stage 1 requirements: the public was invited to comment on the proposal in early 1988. Over a hundred submissions were received in April. These submissions were reviewed by the Ministry for Planning and Environment in May in a report that essentially supported the boating facilities. Most of those opposing the development focused on historic and aesthetic values only. Meetings with interested organisations, an informal community forum, and a workshop with key interest groups were held during preparation of the EES.

A number of newspaper articles published during the EES preparation also highlighted community concerns, particularly the concerns of those involved with the Campaign to Defend the Foreshore, a community action group that organised protest meetings, public exhibitions, and petitions. These concerns included the following issues:

- the potential adverse effects on the beach and foreshore that would result from what was termed a 'radical' transformation of the beach
- discrimination against the owners of small boats
- alienation of the harbour to private interests and loss of existing beach uses
- uncertainty about what the development would entail (Grosse 1990; Saddler 1990).

Alternative views, particularly of members of the Mornington Yacht Club, emphasised the importance of harbour safety. It was also suggested by the President of the Boat Owners Association that 'whether or not a marina is built at Mornington, the work which may affect the condition of Mother's Beach will have to be carried out . . . The marina project seems to be the only way of paying for it' (Tatman 1990).

The EES

In January 1990 the EES (including a summary brochure) was finally completed and published by the consultants on behalf of the Mornington Boating Facilities Working Group (MBFWG 1990). The list of contents is as follows:

- study objectives
- policy context
- existing environmental conditions
- alternative design concepts
- the consultation process
- selection of the preferred development concept
- environmental effects of the preferred development and safeguards against them.

Five alternatives, each at a different location, were considered, as was the 'no go' option. These alternatives were based on five categories of evaluation criteria: level of service, natural values, historic land use or visual character, public access, and finance. Option B (Figure 6.9) in Mornington Harbour was the preferred option, having the least financial costs and impact in relation to the five categories.

Environmental issues raised in relation to Option B included:

- accumulation of sand via inhibition of littoral drift
- change in marine habitats

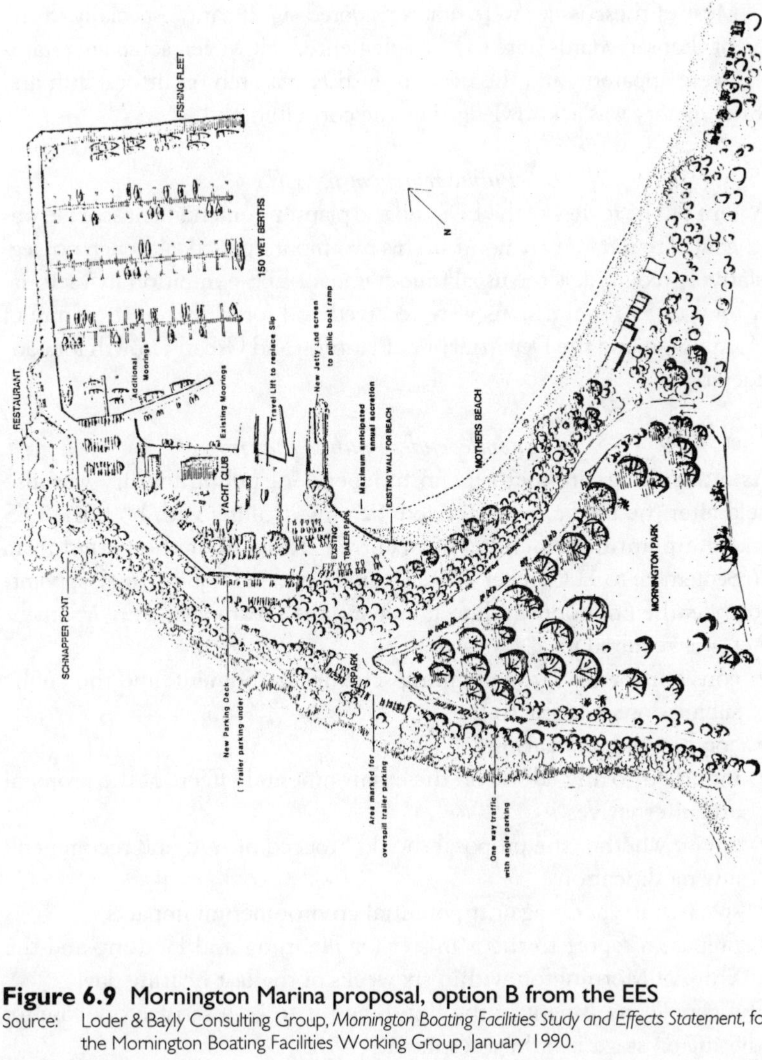

Figure 6.9 Mornington Marina proposal, option B from the EES

Source: Loder & Bayly Consulting Group, *Mornington Boating Facilities Study and Effects Statement*, for the Mornington Boating Facilities Working Group, January 1990.

- loss of benthic fauna (animals or organisms that dwell or feed on the floor of the ocean) as a result of dredging
- increased contaminants in the water (oil, fuels, anti-fouling paints, nutrients, bilge)
- increased traffic
- impacts on historic and landscape values
- impacts on commercial development
- impacts on recreational opportunities.

Most of these issues were not considered significant, especially in the event that safeguards were to be implemented. However, some uncertainties were apparent, and the need for further research on littoral drift and water quality was acknowledged by the consultants.

Public review of the EES

A joint public review of the EES and the planning amendment scheme was scheduled to run for two months. This two-month period, although not legislatively required, is the usual time-frame for EES exhibition in Victoria. A total of 330 submissions were received and forwarded to the Shire of Mornington and the Department of Planning and Urban Growth for consideration.

The independent public inquiry

As often occurs in Victoria, an independent public inquiry was also held after the public review period: hearings dealing with both the EES and the planning amendment occurred throughout the period from 16 September to 11 October 1991. The independent inquiry panel, appointed under the Environment Effects Act and the Planning and Environment Act, was required to:

- consider the EES, the planning-scheme amendment, and the public submissions relating to these
- conduct public hearings
- inquire into and advise on the environmental effects of the proposal and alternatives
- advise whether the proposal should proceed or not, and recommend any modifications
- specify safeguards against potential environmental impacts
- prepare a report to the Minister for Planning and Housing and the Shire of Mornington within six weeks of the last hearing day.

The panel's assessment of the EES found it to be a satisfactory base for advising on stage 1 of the assessment.

This inquiry panel appears to be a significant part of the EIA process in Victoria as it publicly addresses issues that may not have been raised earlier in the EES. In this case, the panel characterised as inadequate the public access provided in relation to an earlier report commissioned by the Mornington Yacht Club in 1987. This earlier Mornington Yacht Club report, prepared by consultants GHD-MacKnight, put forward an alternative development proposal with a larger breakwater and greater boat-holding capacity (300). This proposal was discussed in great detail during the inquiry, despite the fact that the information was not available to the public during the EIA public review period. The panel noted that

> . . . the Mornington Yacht Club's unwillingness to make its information available to the community certainly hindered the Panel and demeaned the debate. Had that information been made available, the level of knowledge in the community would have been higher and the discussions at the Panel Hearing could have led to a better understanding of all issues by all parties.
>
> Panel Report 1991

This omission of information is particularly significant given that the planning-scheme amendment was based on the preferred development concept in the EES (see p. 146), rather than the GHD-MacKnight report. The panel was also highly critical of some comments made by the public at the hearings. These comments were, in their opinion, 'exaggerated, emotional and misleading . . . [and served to] frighten and confuse the public, distort discussion and discredit the conservation movement' (DPD 1993, p. 34).

Because the final panel report summarised public concerns, rather than responding to them separately, it is difficult to get a clear picture of the nature of individual concerns, and it is also hard to determine whether or not all the issues raised by the public were seriously responded to. It is also difficult to ascertain the general level of public support or opposition to the proposal. This characteristic of the inquiry panel, therefore, reduced the transparency of the EIA process, particularly compared with other EIA processes requiring public supplement reports (for example, South Australia) or final EISs (for example, in the Commonwealth jurisdiction). In these jurisdictions, the concerns of the public and the proponent responses to these concerns are more readily accessible to the public.

In relation to the Mornington Marina independent panel report, the key public concerns that can be identified mostly relate to social issues. These included concerns over perceptions that:

- the EES and the planning-scheme amendment arose solely from the Mornington Yacht Club's determination to develop the harbour and become the major tenant
- loose and unquantified provisions in the amendment might be inconsistent with details in the EES
- safe boat-havens already exist 20 km away
- owners of small boats would be 'squeezed out' by the upgraded harbour
- the development would involve unwarranted human impact on the environment
- sewage and oil spills would make the beach unusable
- solid piers would impact on the size and types of fish in the harbour (this concern was expressed by those engaged in fishing activities in the harbour)
- the development would have adverse visual impacts
- there would be a limited market for the proposed restaurant
- an increase in car parking requirements would result
- the development was inadequately financed.

The final panel report was completed on 24 December 1991, within the six-week period required by the terms of reference. The rationale for the harbour was discussed in terms of, among other things, safety, tourism and recreational spin-offs, benefits to the fishing industry, and improved rescue facilities. It was noted that the altered wave conditions, potential sand accumulation, and water quality issues were minor and generally manageable, despite the uncertainties, apparent in the EES, of littoral drift effects. Overall, the panel was 'completely confident that Mornington Harbour can be redeveloped to provide a safe harbour and that any environmental impact can be effectively accommodated in a final concept' (Panel Report 1991, p. 13).

The panel's report put forward 17 recommendations relating to acceptance of the EES as the basis for a stage-1 'in principle' approval. These included recommendations relating to:

- the establishment of a management structure including provisions for monitoring and determination of pollution threshold levels
- the need for further cultural and landscape studies
- changes to the planning-scheme amendment
- specific design guidelines (in relation to, for example, the restaurant, car park, and landscaping).

Of all the development options considered in the EES, the panel supported the preferred development concept, due to its lesser impacts. The panel did, however, note significant advantages in the GHD-MacKnight option (prepared on behalf of the Mornington Yacht Club) given that it had better water-circulation capabilities, greater numbers of boats protected, more convenient fishing fleet mooring, less visual intrusion, greater durability, and greater cost advantages. While care was taken not to recommend any specific proposal for development, the panel did suggest some changes to the planning-scheme amendment. These changes were based on the larger GHD-MacKnight option, and included greater flexibility in zoning to allow for future upgrades and to reduce constraints on the final development. It is interesting to note, however, that this recommendation to allow flexibility in zoning was based on comments by the Mornington Yacht Club, suggesting an element of bias may have been present in decision-making process. In recognition of community concern about larger facilities, the panel also recommended that the state government conduct a study of fluid dynamics and their impacts on the beach.

The assessment report

The next stage of the EIA process required the Minister for Planning to prepare an assessment report. This report was prepared in consultation with the Shire of Mornington, the Minister for Roads and Ports, and the Minister for Conservation and Environment (DPD 1993). While the EIA guidelines (DPD 1995) require that this report be prepared within one month of the inquiry report, the final assessment was delayed by over 18 months until August 1993. Reasons for this delay are not evident, but it should be noted that this one-month period to complete the assessment is not a legislative requirement and is thus discretionary.

The final assessment report by the Minister and the bodies mentioned above was only ten pages and essentially responded to each of the panel's recommendations. Other key recommendations in this final assessment related to:

- shared funding arrangements between the government and a private developer
- establishment of a working group to facilitate stage 2 of the approval process, which required a planning permit, and to provide detailed specifications for the development
- the possibility of foregoing the requirement for an EES at stage 2
- a need for further hydrodynamic modelling in stage 2
- acceptance of the assessment of environmental effects as the basis for approval

- non-acceptance of the restaurant site—alternatively, it was suggested that the restaurant should be integrated into existing Yacht Club facilities.

With the exception of the restaurant and some design points, most of the panel's recommendations were accepted in the assessment report. The assessment report also acknowledged community concerns about the impacts on the value and existing uses of Mornington Harbour, and about the size of the GHD-Macknight proposal as 'too large and threatening to Mother's Beach'. However, even though the assessment report named the preferred development concept in the EES as the recommended option, it also recommended that it be modified, if necessary, in accordance with the panel's recommendations—that is, following the larger form and alignment of the GHD-Macknight plan. They also suggested that any pier extensions should be minimal to maintain aesthetic standards.

Given that the GHD-Macknight proposal was afforded such support in the final inquiry and assessment, but was not publicly available during the public review period, the effectiveness of public participation is questionable in this case. While mechanisms such as the public inquiry are useful to cover any issues ignored earlier, the debate was still hindered by the inadequacy of the information made available to the public in relation to the GHD-Macknight proposal.

In making the decision to approve or refuse the planning-scheme amendment, the Shire of Mornington was required to take the government's assessment report into account. These recommendations were not legally binding. To date, no final decision has been made by the Shire.

WESTERN AUSTRALIA

Formal records for the practice of EIA in Western Australia are available for the period from February 1987 to September 1995 (EPA(WA), pers. comm., 1995). During this period a total of 6630 proposals were referred to the EPA. Of these, 881 were not assessed, 3844 underwent informal or extended informal review, and 743 triggered formal review. During this period there were 543 CERs, 139 PERs, and 20 ERMPS. In relation to appeals, 773 were appeals against the level of assessment required, 1130 were appeals against the EPA(WA) assessment reports, and 11 were appeals against Ministerial conditions. One particular proposal was the subject of 252 appeals over a period of one year (1987/88).

The Quicklime project CER

The proposal

In 1990, in light of growing demands for quicklime in Western Australia, the company Swan Portland Cement Limited (Swan) proposed to construct a new quicklime plant in the Nowergup area. This plant was to have a capacity of 230 000 tons per year and a life span of 30 years. Limestone was to be quarried from adjacent mineral leases and freehold land at a rate of 450 000 tons, with a possible increase to 900 000 or more tons each year (resulting in production of 46 000 tons per year of quicklime). The proponent noted that such an increase would require an additional assessment by the Environment Protection Authority (EPA(WA)), and this additional assessment occurred later in the EIA process. The conceptual layout of the revised plant proposal includes the following:

- office and workshop areas
- a dump pad
- a crushing and screening station
- a transfer station
- limestone storage facilities
- a mill building
- a precipitator
- a surge bin
- a compressor house
- transformers
- a silica-reject silo and cooling tower.

Transport of quicklime would be via road bulk-tankers, with a possible movement of 60 vehicles per day. A 12-month time period was estimated for construction, and its operations were planned to continue for 24 hours per day.

A CER is required

In August 1990 the proposal was referred to EPA(WA), as required by section 38 of the Environmental Protection Act. Of three formal levels of assessment (CER, PER, ERMP), it was the CER that the EPA(WA) required. The EPA(WA) then issued project-specific guidelines pursuant to section 40(3) of the Environment Protection Act. The CER is the lowest form of formal assessment in Western Australia, being reserved for projects with relatively manageable and localised impacts, and localised public concern. No other jurisdiction has this third level of formal assessment, with the possible exception of Tasmania, and the recently amended Development Act in South Australia (Development Report).

The CER (Kinhill Engineers Pty Ltd 1991) was prepared by consultants Kinhill Engineers on behalf of Swan in May 1991, approximately eight months after the initial referral to the EPA(WA). Relative to the other EIA examples summarised, the CER document is short and concise, reflecting the lower level of assessment. Its contents reflect the EPA(WA) guidelines and include sections on the following issues:

- project justification
- evaluation of alternatives (site, mining strategy, process technology, transport options, the 'no go' option)
- project description
- the existing environment
- environmental impacts and the management of these (including decommissioning and rehabilitation)
- monitoring
- public consultation
- commitments to protect the environment.

Environmental issues identified in the CER include:

- localised social impacts (employment, transport, aesthetics, noise, and dust)
- impacts on groundwater quality and quantity
- atmospheric emissions
- impacts from solid and liquid waste disposal
- impacts on biota (vegetation loss, noise, dust).

The consideration of alternatives was notably more explicit in this CER than in the other case studies discussed above—for example, the ERMP prepared in relation to the Mt Lyell Mining redevelopment proposal in Tasmania.

Public review during preparation of the CER

Although not required by the Act, an informal public meeting was held on 12 January 1991 to assist in the preparation of the CER. Approximately 30 people attended, and most of the issues raised related to traffic issues and noise. Other issues included dust problems, visual impacts, atmospheric emissions, future land-use plans, rehabilitation, and water sources.

As required by section 40(4) of the Act, the CER was released for a public review period of four weeks, commencing on 9 May 1991 and ending on 6 June 1991 (although it should be noted that the actual time-frame is not specified in the Act). A total of 25 submissions were received: 18 from the public (including one group-submission of 29 people) and seven from government agencies. These were summarised by EPA(WA) and forwarded to the proponent, Swan, for its response. The issues and proponent's response were outlined in the EPA(WA) assessment report.

Issues raised in the EPA(WA)'s assessment report of the Quicklime proposal

- relocation of plant site
- impacts on water supply and quality
- impacts on flora and fauna (rare species in the region)
- land clearance
- noise
- dust
- gas emissions
- social impacts arising from the close proximity of plant to local residents

The formal public review stage had a considerable influence on the project in that public concerns caused the proponent to choose an alternative plant site on 17 July 1991, following liaison with EPA(WA), the Social Impact Unit (previously attached to the Department of State Development, but now disbanded), and the Department of State Development. Another informal public meeting was held to discuss the new plan. Forty people attended, the majority of whom were 'satisfied that their concerns had been adequately addressed and that the alternative site was a better location for the plant' (EPA(WA) 1991, p. 8). The EPA(WA) was formally notified of the final changes on 2 October 1991.

This approach to the public process was highly commended by EPA(WA), which stated that

> The proponent's public consultation programme has been sound and comprehensive. The Authority commends Swan for its efforts in identifying the major issues of concern to local residents, and for making genuine efforts to resolve these issues, particularly by being prepared to relocate the proposed quicklime plant to a more acceptable site.
>
> EPA(WA) 1991, p. 8

The assessment report

Following the public review, EPA(WA) produced an assessment report (EPA(WA) 1991) pursuant to section 44 of the Environmental Protection Act. Again, this was a relatively concise document, reflecting the lower level of assessment. Overall, the assessment report characterised the proposal as 'environmentally acceptable' and described the impacts as manageable. The report recommended that the project should proceed. Four additional recommendations were also made relating to

- preparation of a management program for the species *Eucalyptus argutifolia*
- minimisation of noise
- preparation of an environmental management program to monitor and manage impacts
- control of pollution via a licence requirement under the Environmental Protection Act.

The decision

Based on the recommendations of the EPA(WA), the Minister for Environment approved the project, subject to legally binding conditions, on 8 June 1992. The final decision was made approximately seven months after the recommendations of the EPA(WA). While the reasons for the delay are unclear (it may be normal practice), it does appear to be a long time-frame relative to Tasmania's Mt Lyell redevelopment, where the *whole* assessment process took six months. This difference is significant given that the assessment level in Western Australia was the lowest (localised impacts), while in Tasmania it was set at the highest level (proposal of state significance).

The project is revised

However, the final decision did not, in this case, signal the end of the process. Over a year after the approval, the proponent decided to increase the capacity of the plant site to 430 000 tons per year, as indicated in the original CER (Figure 6.10 shows the conceptual layout of the revised proposal). Preliminary discussions between the proponent and the EPA(WA), in May and September 1993, resulted in the EPA requiring the proponent to refer the proposal back to the EPA for investigation, pursuant to section 46 of the Environmental Protection Act (this section relates to the amendment of conditions and procedures). The formal referral document, produced by Kinhill Engineers Pty Ltd (1993) on behalf of the proponent aimed to assess the incremental effects resulting from the revised proposal, and to assist

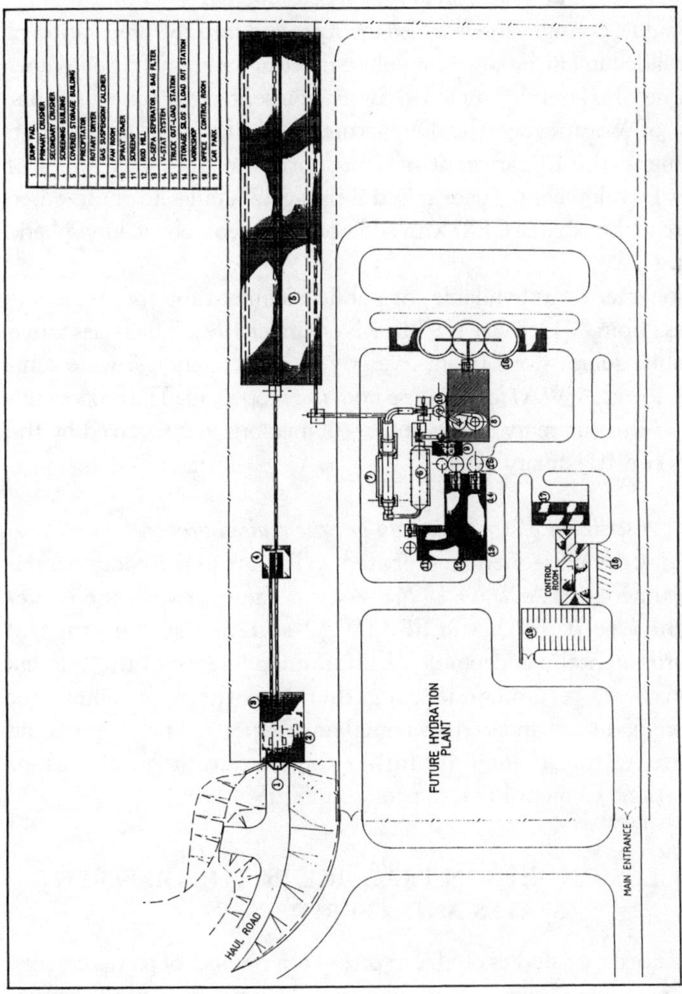

Figure 6.10 Nowergup quicklime plant proposal (revised version)
Source: Kinhill Engineers Pty Ltd 1993

the EPA(WA) in determining the appropriate level of assessment. However, no level of assessment was set, and the referral document was used as the basis for the final assessment by the EPA(WA). The contents of the referral document were similar to the earlier CER.

Public review of the revised proposal

Informal public consultation was also incorporated into the preparation of the referral document by way of a public meeting held on 18 September 1993. Those who attended included Swan representatives, nine residents, the City of Wanneroo, the Department of Planning and Urban Development, the Department of Transport, and the Department of Resources Development. Issues raised were again similar to those raised in response to the original CER, with additional concerns about atmospheric emissions.

The document was available for public comment for approximately three weeks from 20 October 1993 to 9 November 1993. The issues raised in the public submissions (from five government agencies) were summarised by the EPA(WA) for response from the proponent. The proponent's response to this summary of the public submissions was received by the EPA(WA) on 10 January 1994.

Assessment of, and decision on, the revised proposal

The EPA(WA) subsequently prepared a 'Report and Recommendations' document (EPA(WA) 1994). As was the case with the earlier assessment (see p. 156), the EPA(WA) stated that the proposal was 'environmentally acceptable'. Minor amendments to the original conditions were recommended, together with the requirement for compliance auditing, and regular compliance reports. The proposal did not eventuate for commercial rather than environmental reasons (Swan Portland Cement Pty Ltd, pers. comm., 1996).

CONCLUSION: EIA IN PRACTICE IN THE DIFFERENT STATES AND TERRITORIES

From the above case studies of EIA in practice, a number of points emerge. First, the passage of a project through the EIA process is not always as clear cut as would appear from looking at the legislation alone, particularly where more than one EIA jurisdiction is involved. For example, the Magnetic Quay project (formerly Magnetic Keys) involved both state and Commonwealth legislation. The project had requirements for three separate forms of EIA (IAS, PER, and EIS), with the first two of these deemed inadequate and requiring second attempts. Although this produced a massive amount of EIA documentation, the project was approved (with financial guarantees through insurance), and earthworks commenced before financial problems halted the project and left the impact of environmental degradation adjacent to a National Park and World Heritage area. Amelioration was not possible because the insurance had lapsed and the financial guarantees were longer in place. In other cases, such as Hindmarsh Island, the EIA process appeared simple, documentation was satisfactory, and the process was completed in record time. However, this did not prevent the subsequent controversy that has stalled the project.

Second, the public involvement in EIA projects can also become quite complicated and controversial. The Sydney Harbour Tunnel EIS had an extended period of public review and attracted significant public concern prior to its approval. This prompted the then Liberal opposition to promise a public inquiry if it was elected to power. The Liberals came to power but held no inquiry. Instead, the local council held its own public inquiry and identified a number of major concerns, including the financial viability of the project. The council failed to stop the project, and as predicted, significant financial losses have been incurred. In the Hindmarsh Island case, there was a major difference between the impact of formal public involvement (as required by the EIA legislation) and significant informal or ad hoc public involvement associated with the EIA process. This raises questions about the effectiveness of public involvement, and the correct identification of stakeholders.

Third, the separate state and territory EIA provisions become complicated when there are overlapping jurisdictions. The EIS for the Tindal Airbase project was constrained in its investigations of issues such as infrastructure expansion because these issues were to be examined separately under Commonwealth legislation once additional funding was made available. The EIS for the Magnetic Quay project was also constrained by Commonwealth jurisdiction within the Great Barrier Reef Marine Park. To a large extent, many problems of multiple jurisdictions have been addressed in recent cooperative arrangements (see Chapter 8).

Fourth, it is clear that there are numerous other influences on the smooth passage of the EIA process that cannot be gleaned from the legislation. The timing of different elements of the process varies considerably—for example, the length of time taken to complete an EIS or the time taken for proponents to respond to comments. In a South Australian study Harvey (1994a) demonstrated that over 70 per cent of the time taken to complete the whole EIA process can be attributed to proponents and their consultants rather than to governments or government instrumentalities. Other important influences on the process relate to the nature of the project and its location; the proponent, and their EIA consultants; liaison between the consultant, proponent, and government; political influences; and political decision-making. In relation to political influences, the EIA process is often influenced by either associated legislation, which varies between jurisdictions, or by legislation created specifically for EIA projects—for example, the Sydney Harbour Tunnel or the Hindmarsh Island bridge.

Fifth, the review or assessment process is conducted mostly as a government 'in-house' operation. This is similar to many other countries, with some notable exceptions such as Canada, where there is an independent review committee. In Australia the role of watchdog is primarily carried out by government, with the public playing a lesser role in this respect. There are a number of exceptions to this, many of which reflect political intervention in development proposals—for example, the South Australian marinas. In other cases, the public are not constrained by the formal period for public involvement in the EIA process and adopt a watchdog role, even once the process has been completed—for example, in the case of the Hindmarsh Island bridge.

Sixth, it is clear that the EIA process introduces a cost and often delays for development projects. For responsible proponents who would normally consider environmental factors and undertake proper consultation, the EIA process does not represent a significant burden. For other proponents, the imposition of adequate environmental investigations may actually save costs in the long run. However, the practice of public and government consultation varies significantly with each case and, if not handled carefully, can introduce delays and cause inefficiencies in the process.

Finally, the case studies demonstrate some of the inadequacies of the EIA process as a mechanism for dealing with the complexity of environmental issues, some of which are specific to the project and EIA jurisdiction. For this reason, a true understanding of EIA within any one jurisdiction requires a detailed investigation of the legislative and administrative procedures in addition to a careful study of what happens in practice. This chapter provides only a snapshot of some of these intricacies. In Chapter 7 Commonwealth EIA legislation is considered separately, as it can apply, and overlap, with any of the state and territory EIA processes described above.

7

Australian Commonwealth EIA legislation and practice

Unlike EIA legislation in the states and territories, Commonwealth EIA legislation has been in existence for a longer time (since 1974), and it is applicable to the type of project or approval rather than to a regional area of jurisdiction. It is also possible for the Commonwealth to require an EIA for a project that is simultaneously subject to individual state or territory EIA requirements. For these reasons the Commonwealth EIA process is discussed here in a separate chapter.

Within two years of the introduction of the NEPA, in the United States (in 1970), the Australian Environment Council had established a working party on environmental impact assessment procedures. By December 1974 the Commonwealth had introduced EIA legislation, the *Environment Protection (Impact of Proposals) Act 1974*, and Administrative Procedures established under section 6 of the Act. The EIA process under this Act was essentially modelled around EIA under the NEPA but was more discretionary to minimise potential delays from litigation. The Administrative Procedures provide most of the guidance for the process, although there is some debate regarding their legal status.

The Environment Protection (Impact of Proposals) Act falls into separate parts dealing with definitions, the object and application of the Act, the preparation of separate Administrative Procedures for EIA, and provisions for separate public environmental inquiries. The Administrative Procedures came into effect in July 1975, but as with the Act, they have been amended a number of times (most recently in 1995). For example, an amendment in 1987 introduced the public environment report (PER) as an intermediate level of assessment.

The Environment Protection (Impact of Proposals) Act allows for the Minister for Environment to require an EIS for any proposal likely to have a significant or controversial impact on the environment. This applies only to projects conducted or funded by the Commonwealth government or requiring its approval. The Act has been administered under the 'environment' portfolio, although this portfolio has had three different names over the last decade. It used to be called the Department of Arts, Sport, the Environment, Tourism, and Territories (DASETT). Since 1991 the Act has been the responsibility of the Commonwealth Environmental Protection Agency (CEPA), created in August 1991 as a separate organisation within the Department of Arts Sport, the Environment, and Territories (DASET). However, in 1993 the Arts was dropped from DASET so that the department became the Department of Environment, Sport, and Territories (DEST). Additional administrative changes were made in 1996, when CEPA became part of the Environment Protection Group within 'Environment Australia', an administrative unit containing most of the Commonwealth's environment-related divisions. Within the Environment Protection Group is the Environmental Assessment Division, which is responsible for EIA. Joint assessment arrangements between states and the Northern Territory are provided for under paragraph 12.2.1 of the Commonwealth Administrative Procedures.

SCREENING, SCOPING, AND LEVELS OF ASSESSMENT

In the Act, the term 'environment' is defined as 'all aspects of the surroundings of human beings' and includes the natural environment, the built environment, and socio-economic factors. The definition covers such factors as air, water, soils, flora, fauna, buildings, roads, employment, housing, and recreational facilities.

Generally, proposals subject to the Commonwealth EIA legislation fall into one or more of the following categories:

- activities and projects carried out by Commonwealth departments and authorities—for example, defence projects, railways, national highways, airports, postal and telecommunication facilities, developments on Commonwealth land

- projects funded by Commonwealth grants—for example, bicentennial road programs, international sports facilities, special grants
- projects requiring Commonwealth approvals for exports—for example, uranium, coal, iron ore, mineral sands, aluminium, oil, gas, woodchips
- foreign investment projects—for example, mining, manufacturing, real estate, tourist development.

The Environment Protection (Impact of Proposals) Act only applies to matters under Commonwealth jurisdiction that may affect the environment 'to a significant extent' (section 5), although it has the potential to extend beyond projects. It may include, for example, the formulation of proposals, the negotiation of agreements, the making of decisions and recommendations, and the incurring of expenditure. As Wood (1995) states, the legislation was clearly intended to apply to policies and other Commonwealth actions, as well as to projects. However, in practice it has focused mainly on projects.

The first 'triggering' step rests with the action Minister (not necessarily the Minister for Environment) who is the Minister of the department responsible for the proposed development or action. Triggering of the process is discretionary, and if the action Minister does not consider it appropriate to refer the proposed action to the Minister for Environment, then the Environment Protection (Impact of Proposals) Act is not invoked. If a matter falls under the scope of the Act, the action Minister designates the proponent who must supply the Minister or DEST with further information, usually in the form of a notice of intention. The notice of intention must included the following information:

- a summary of preliminary plans and any alternatives that have been considered
- a description of the environment likely to be affected by the proposed action or by any alternatives
- an assessment of the potential environmental impacts of the proposed action or any alternatives, including any beneficial impacts
- a description of any environmental protection safeguards or standards intended to be implemented in relation to the proposal
- a statement of any environmental impact investigations or studies intended to be carried out in relation to the proposed action (Administrative Procedures, section 2.2).

According to Wood's (1995) review of seven EIA jurisdictions, Australia is the only jurisdiction (out of the seven he reviewed) that is unsatisfactory in terms of its screening procedures. Woods attributes this to the lack of control by the Minister for Environment and to uncertainty regarding criteria.

The Environment Protection (Impact of Proposals) Act contains four levels of assessment:

1. examination but no EIS or PER
2. assessment after preparation, and public review, of PER
3. assessment after preparation and public review of EIS
4. commission of inquiry.

The first level relates to the situation where no PER or EIS is required and requires only recommendations from DEST. The next two levels, the PER (introduced in 1987) and the EIS, are both the responsibility of the proponent. The fourth level, the commission of inquiry, is a government responsibility. For a time there was overlap between this level of EIA and the now defunct Resource Assessment Commission, which had its own Act and had, by 1993, completed three inquiries.

Once the action Minister has referred the matter and has required a notice of intention to be prepared, it is then up to the Minister for Environment to require either an EIS or a PER to be prepared. However, the Department (DEST) can itself determine that an EIS or a PER is not required. In order to arrive at such a decision DEST or the Minister for Environment must consider the criteria contained in paragraph 3.1.2 of the Administrative Procedures.

Criteria for determining the need for Commonwealth PER or EIS

a) whether and to what extent the proposed action would result in:
   - a substantial environmental effect on a community
   - the transformation of a substantial area
   - a substantial impact on the eco-systems of an area
   - a significant diminution of aesthetic, recreational, scientific or other environmental quality, or value of an area
   - an adverse effect on an area, or structure, that has an aesthetic, anthropologic, archaeological, architectural, cultural, historical, scientific or social significance or other special value for the present or future generations
   - the endangering, or further endangering, of any species of fauna or flora

> - important long-term effects on the environment
> - the degradation of environmental quality
> - curtailment of any beneficial uses of the environment
> - the pollution of the environment
> - environmental problems associated with waste disposal
> - increased demands on natural resources that are likely to be in short supply; and
>
> b) any environmental assessment action taken relevant to the proposed action by any State or the Northern Territory or authority of a State or the Northern Territory.
>
> Administrative Procedures, section 3.1.2

Guidelines to assist in identifying what is meant by 'environmental significance' have also been formulated by CEPA in the form of a memoranda of understanding (Münchenberg, 1994a).

Matters to be addressed in the EIA, in addition to matters outlined in the Administrative Procedures (paragraphs 4.1 and 4.2), are discussed in meetings between the proponent, Environment Australia, and any other government agency involved. Following consultations, the Minister may also outline the matters to be addressed. Public involvement at this stage is not specified in the Procedures, although it may occur in practice (CEPA 1995).

The Administrative Procedures outline a number of procedural steps (see Figure 7.1, p. 167). The main steps are as follows:

- the proponent of a proposed action is designated—this could be a Commonwealth department or authority, or a private sector body
- the proponent provides preliminary documentation, usually a notice of intent describing the proposal, the environment to be affected, the expected beneficial and adverse impacts, any alternatives to the project, and the proposed environment protection measures
- the information is assessed by DEST, which administers the Act through CEPA
- DEST advises as to whether or not an EIS or PER is required
- Environment Australia prepares project-specific guidelines
- the documentation is prepared and made available for public review
- the proponent revises the EIS (this does not happen in relation to PERs)

- DEST makes an assessment
- the Minister for Environment makes recommendations, comments or suggestions.

As noted above, Environment Australia may determine that an assessment is not required, in which case the Minister for Environment must give reasons in a public statement, or alternatively must refer the decision to the action Minister, who may consult with government agencies for final determination. Criteria for making this decision are specified in paragraphs 3.1.2 and 7.2 of the Administrative Procedures and are publicly available. The Minister may decide an EIS or PER is not warranted if it is not considered to be in the public interest and, under paragraph 11.1 of the Administrative Procedures, may also exempt any actions from the requirements of the Administrative Procedures if these actions are prejudicial to national security or to the interests of Australia, if they are contrary to public interest, or if they adversely affect commercial or other interests. It is interesting that the recent 1995 amendments to the Administrative Procedures allow the Minister for Environment to enable decisions to proceed while the assessment is still being undertaken (for example, in relation to export approvals), although this decision must be consistent with public interest and the views of the action Minister.

EIS PREPARATION AND PUBLIC PARTICIPATION

An EIS or a PER must contain a clear and concise summary of the content and presentation of the report, and must minimise the production costs so as to make it available to the public as widely as possible. The requirements regarding the contents of the EIS and PER differ slightly (see Table 7.1).

Once the EIS or PER has been prepared, it is placed on public exhibition for a minimum period of 28 days, although the EIS period is often extended. In addition, the proponent must seek comments from the government agencies specified in paragraph 6.4 of the Commonwealth Administrative Procedures.

Following Ministerial direction, Environment Australia may also hold round-table discussions with the public and the proponent within 28 days of the public review period, although Kinhill Engineers (1994) states that this has not yet formally occurred in practice. Once all comments have been submitted to the proponent, the proponent must respond, in a final EIS, to:

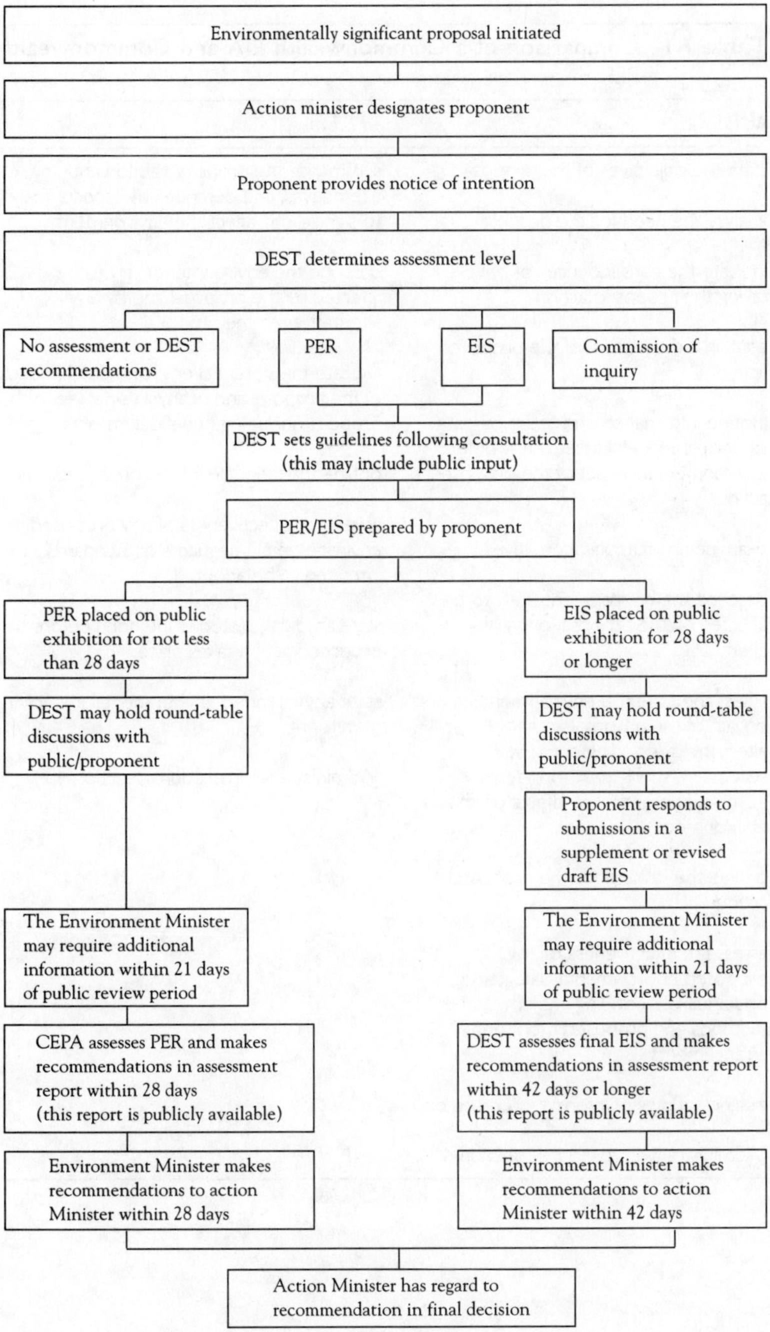

Figure 7.1 Commonwealth EIA process

Table 7.1 Comparison of a Commonwealth EIA and Commonwealth PER

An EIS must	*A PER must*
state the objectives of the proposal	summarise the proposed action and any alternatives, and describe any aspects likely to significantly affect the environment
analyse the need for the proposal	describe the environment likely to be affected by the proposal and by any alternatives to the proposal
indicate the consequences of not taking the proposed action	indicate the potential environmental impact of the proposal and of any alternatives to the proposal, including beneficial impacts
contain a description of the proposed action	outline the rationale for the proposed action
include information and technical data to permit a careful assessment of the environmental impact of the proposed action	assess the effectiveness of any proposed environmental safeguards or standards intended to be adopted
examine any feasible alternatives	state any further studies intended before the proposed action takes place
describe the environment likely to be affected by the proposal or by any alternatives	state any intended environmental impact monitoring
assess the potential environmental impact of the proposal and any alternatives, including primary, secondary, short-term, long-term adverse, and beneficial effects on the environment	(Administrative Procedures, section 4.2)
outline the rational for the proposed action	
assess the effectiveness of any proposed environmental safeguards or standards, including implementation methods and proposed monitoring arrangements	
outline sources of information relied on	
(Administrative Procedures, section 4.1)	

- written public comments
- any Environment Australia report arising out of round-table discussions
- any inquiry reports (this is not required for a PER).

However, the decision to release the documents for public exhibition is discretionary (paragraph 6.2.3 of the Administrative Procedures).

ASSESSMENT, DECISION-MAKING, MONITORING, AUDITING, AND APPEAL RIGHTS

Within 42 and 28 days, respectively, of the final EIS and PER being completed, Environment Australia examines the documents, together with public submissions on the PER, and then reports to the Minister for Environment in an assessment report. The Minister for Environment may require additional information, within 21 days, from either the EIS or the PER. Subsequently, the Minister for Environment makes comments or recommendations, as well as any conditions that should be required as part of an approval, to the action Minister. With the exception of confidential information, this report is publicly available. Advice is non-binding and the Minister responsible for the decision need only take 'into account' the assessment (section 8, par. 9.5 of the Administrative Procedures) when making a decision.

There appears to be no provision for appeal on the decision (Kinhill 1994). Münchenberg (1994a) summarises four barriers to litigation, including costs, difficulties in proving standing, the uncertainties regarding the legal status of the Administrative Procedures, and the discretion inherent in much of the process. At the same time, Münchenberg states that nothing in the Administrative Procedures completely bars third-party litigation. Monitoring and review of the proposed action, particularly in terms of environmental safeguards, are catered for under paragraph 10.1.1 of the Administrative Procedures, and any reports produced are publicly available. A PER may also be required, if necessary, following review of the proposed action.

EIA IN PRACTICE UNDER COMMONWEALTH LEGISLATION

Since 1974 over 2600 proposals have been considered environmentally significant and forwarded for assessment. Of these, 164 had EIS requirements (as of July 1997), 30 required PERs, and five triggered inquiries, while the rest were assessed with recommendations by DEST (now Environment Australia) (CEPA 1995). The inquiries include the Ranger Inquiry in the Northern Territory (uranium mining), the Fraser Island Inquiry in Queensland (sandmining), the Shoalwater Bay Inquiry in Queensland (resource use), the Ulladulla Inquiry in New South Wales (transmission station), and the East Coast Armaments Complex in Victoria (CEPA 1995).

The Sydney third runway project

This project is interesting in that it is one of the largest Australian EISs to be produced (720 pages), and it was conducted entirely under the Commonwealth EIA legislation. There was, however, a High Court challenge by Botany Bay Council requesting that the proposal should also have fallen under New South Wales jurisdiction. This challenge was rejected, although subsequently there was significant cooperation between the Commonwealth and the New South Wales government. The Sydney third runway project is also interesting because the project has been subject to intense public controversy, even after the final decision. This suggests that the public concerns were not taken seriously in the process, and that the government and the private interests involved perhaps saw the EIA process as an administrative hurdle to be overcome after the decision had already been made.

On 22 March 1989 the Prime Minister and the Minister for Transport and Communications announced plans to construct a third runway parallel to the existing north-south runway at Kingsford Smith Airport in Sydney. The aim was to accommodate future passenger demand in the medium term to the year 2010 (FAC 1990; DASETT 1991). The proposal (Figures 7.2 and 7.3, pp. 170–1) involved:

- reclamation of 170 ha of land, by dredging 14 million m^3 of sand in Botany Bay
- construction of a 2400 m runway
- breakwater alteration
- diversion of a stream outlet
- relocation of the airport control tower
- new fire-fighting and rescue facilities
- establishment of corporate aviation facilities (FAC 1990; Blumer 1993).

In its broader context, the proposal forms part of a three-part integrated strategy proposed in April 1988, which includes the implementation of short-term traffic management measures until new runway facilities become available, and long-term staged development of Sydney's second airport (FAC 1990).

On 21 March 1989, prior to the announcement, Cabinet had determined the need for an EIS (DASETT 1991). Given that the project fell within Commonwealth jurisdiction, the Commonwealth Environment Protection (Impact of Proposals) Act was triggered, and the Federal Airports Commission was nominated as the proponent. In this case, the then Minister for Shipping and Aviation Support was the action Minister (FAC 1991). There was also a degree of cooperation with the New South Wales EIA process in the 'spirit' of the 1983 non-legislative agreement outlining 'procedural guidelines for environmental assessment involving the commonwealth and the state of NSW' (FAC 1990). This was despite the fact that the proposal did not fall under the jurisdiction of the New South Wales Environmental Planning and Assessment Act.

There was, however, a degree of controversy about restricting the scope of assessment to Commonwealth procedures. This was raised in the courts by Botany Municipal Council (Hill 1992). The Council asserted that the dredging component of the proposal was outside the jurisdiction of the Federal Airports Commission. It was also argued that the Federal Airports Commission was in fact a 'determining authority' under the New South Wales process, and should thus comply with EIA processes under Part 5 of the New South Wales Act. The High Court rejected these arguments and the proposal remained under Commonwealth jurisdiction (Hill 1992).

A notice of intention was prepared by the Federal Airports Commission and received by DASETT on 7 July 1989. Shortly afterwards, a formal requirement for an EIS was made by the then Minister for Arts, Sport, the Environment, Tourism and Territories (Senator Graham Richardson) on 1 September 1989. Note that draft guidelines had been formulated prior to this decision (25 August 1989) by DASETT in consultation with the Federal Airports Commission, the Department of Transport and Communications, and the New South Wales Department of Planning. Although the release of guidelines for public comment is not required under the Commonwealth Act or the New South Wales process, these guidelines were released for public comment for a one-month period (running from the date of advertisement of the guidelines on 26 August). Final guidelines were issued on 13 November 1989. They incorporated consideration of over 250 public submissions.

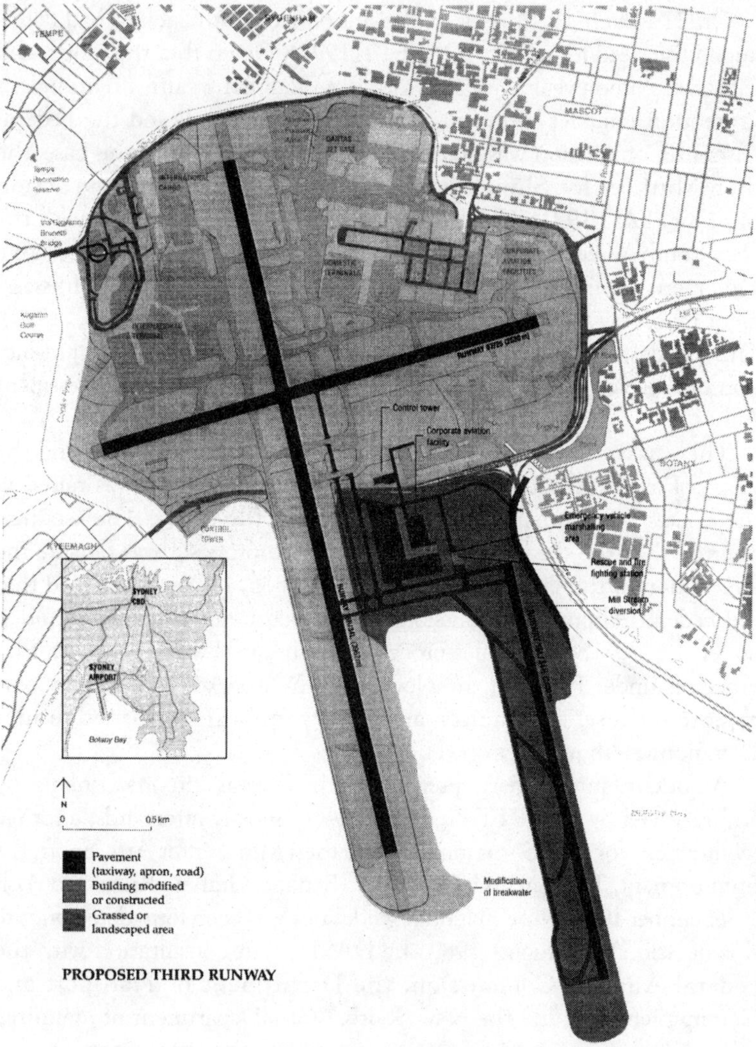

Figure 7.2 Sydney third runway proposal
Source: FAC 1990

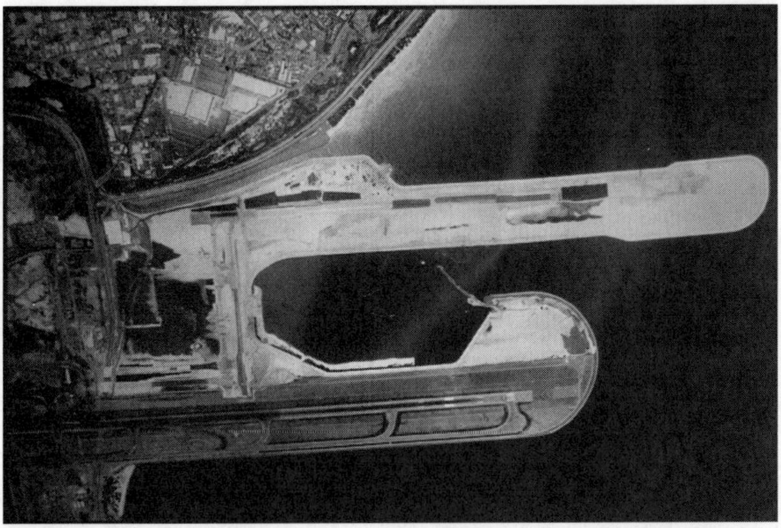

Figure 7.3 Photograph of Sydney third runway site
Source: supplied by Federal Airports Corporation

The draft EIS was completed just over one year later, in September 1990, by consultants Kinhill Engineers Proprietary Limited on behalf of the Federal Airports Commission. Ballantyne (1990) reports that this draft EIS, which is 720 pages long, has been characterised as the most comprehensive EIS in Australia.

Environmental issues raised in the Sydney third runway EIS

- potential impacts on Aboriginal and European heritage
- changes in surface hydrology
- impacts on flora and fauna
- loss of saltmarsh and other avifauna habitats
- bird-strike potential
- impacts on landscape and visual environment
- impacts of extracting fill (for example, via offshore dredging)
- changes in coastal hydrodynamics, reduction of wave energy, and minor beach re-alignment resulting, in part, from dredging

- impacts on marine ecology (water quality, sediments, seagrasses, fish) from dredging and reclamation—for example, suspended solids, pollution
- pollution control and waste disposal
- site access
- traffic impacts
- restrictions on the use of adjacent land
- social and economic impacts—for example, traffic, dust from construction, aircraft noise affecting health, residential satisfaction, recreation and community facilities, effects on property values
- potential hazards—for example, fuel spills, collisions, fires, explosions
- impacts on air quality from the following: aircraft operations, surface traffic, aircraft maintenance, fuel storage and distribution
- effects on the volume of surface traffic

Although review of the draft EIS is again not a legislative requirement, this draft EIS was reviewed by DASETT in terms of its adequacy for public release (Jenkins 1993). It is interesting to note that, prior to public release, the then responsible Minister, Senator Bob Collins, noted that none of the issues raised were 'big enough [to] . . . preclude the development of the third runway' (Ballantyne 1990). The general manager of the Federal Airports Corporation described the EIS as 'impartial, thoroughly professional and painstaking' (Ballantyne 1990). But, in stark contrast, 14 local municipalities were prepared to commit $500 000 to an independent review to challenge the methodology and assumptions made by Kinhill in the EIS (Bita 1990).

Due to strong and controversial public interest, the EIS was released for public exhibition for three months, from 20 September to 20 December 1990, rather than the usual 28 days. This was extended for 13 days after the review period, and the public exhibition period finally ended on 3 January 1991, although some submissions were accepted after this date. A large number of submissions were received (1897), together with 2283 form letters, 5804 postcards, and 3850 signatures on petitions. Public consultation with invited community groups was undertaken by DASETT and the Federal Airports Commission. This participation process occurred between 22 and 23 April 1991.

Approximately 87 per cent of the public submissions opposed the proposal. Approximately 8 per cent voiced support, mainly because of the potential adverse impacts on business and tourism if the runway were not to proceed. Most (68 per cent) of those opposing the runway were situated in the northern noise-affected zone. Overall, there were 14 000 separate issues raised during the public review period, most of which referred to Part B of the EIS (the analysis of need and alternatives—34 per cent of submissions) and Part F (impacts from project operation—45 per cent) (FAC 1991). The issues receiving the most comments were the impacts of aircraft noise and the associated effects on health, education facilities, property values, quality of life, monitoring, prediction methods, and compensation.

Apart from noise effects, the consideration of alternatives in relation to costing, runway layout, and adequacy of the evaluation process were also frequently addressed, as were the potential hazards of the proposal (for example, safeguards, increased aircraft crashes, and increased bird-strike risk). Other issues raised in the submissions related to:

- the adequacy of both the public consultation process and the review period, given the complexity of the proposal
- impacts on land use and zoning
- impacts on heritage
- changes in the water table
- insufficient data on air-traffic control
- possible bias
- complaints about the voluminous and complex nature of the EIS
- concern about the practice of requiring proponents to prepare the EIS
- general concerns about the EIS process
- calls for a public inquiry, although this idea was later rejected by the Minister for the Environment (summarised in DASETT 1991; FAC 1991).

In addition to the normal public review period, the Federal Airports Commission also initiated an extensive public community consultation program in October 1989. This was claimed to be one of the most extensive programs for an EIS in Australia, and entailed:

- technical workshops with invited guests
- establishment of four community access-centres (these received over 20 000 visitors)

- a mobile access-centre
- telephone information access (550 calls were received)
- distribution of 85 000 newsletters in various languages
- two community-attitude telephone surveys (although their validity was questioned by the public and DASETT)
- noise workshops to outline methodology for assessing noise
- ongoing liaison with local, state, and Commonwealth government authorities, including a meeting for 20 municipalities during preparation of the draft EIS, liaison with special interest groups, and request for, and receipt of, 1284 public submissions from October 1989 to June 1990 (these were taken into account during preparation of the draft EIS and are summarised in tables attached to the draft EIS)
- an offer of Commonwealth government funding ($100 000) to a number of community groups to facilitate participation in the EIA process.

Following completion of the public review period, the proponent summarised and responded to the issues in a supplement, which was completed in September 1991 (two volumes), and received by DASETT on 2 October 1991. In doing so they were required by DASETT to take into account 'matters concerning the content of the EIS' raised in the April meetings (see p. 175) held during the formal period of public review. During the preparation of the supplement, the Federal Airports Commission also consulted with DASETT, and with Commonwealth and New South Wales government agencies (the Civil Aviation Authority, the Commonwealth Department of Transport and Communications, the New South Wales Department of Planning, and the Maritime Services Board).

The supplement contained a modified runway layout, with a slight increase in separation between the existing and proposed runways and associated changes in layout. Additional information was also provided on:

- alternatives
- dredging
- an environmental management plan (the terms of reference were prepared by the Federal Airports Commission, the New South Wales Department of Planning, the State Pollution Control Commissions, and the Department of State Development)
- construction strategies
- air quality

- use of the east–west runway
- bird-strike and noise-management strategies.

The proponent is usually required to forward a copy of the supplement to everyone who wrote a submission, but given the large number of responses, the Minister for Environment exempted the Federal Airports Commission from this process. However, letters of offers for copies were sent to those who had made submissions and the supplement was available for public sale.

The EIS and public submissions were subsequently examined in an assessment report prepared by the Environment Assessment Branch of DASETT within the required 42-day time-frame (by 8 November 1991). Comments and support for the proposal from the New South Wales Department of Planning were also incorporated into an appendix of the report. It was suggested that the associated environmental impacts could be adequately managed with proper conditions and enforcement, and that there would, in fact, be a reduction in noise impacts for 15 000 residents.

Key assessment issues raised by the Environment Assessment Branch in this report included:

- the adequacy of the demonstration of the need for the runway (this demonstration was deemed to be successful)
- consideration of alternatives (need for trade-off)
- environmental management plans
- air quality
- hazards and risk
- sediment control
- water quality
- coastal hydrodynamics
- marine impacts
- migratory waders (a type of bird)
- surface transport and tourism.

The effect of aircraft noise was considered to be the main issue. Consideration of noise issues included discussion of the benefits to those in the east and west (whose exposure to the flight paths would be changed), consideration of the increased adverse impacts on northern residents, and discussion of the uncertainties of assessing the magnitude of the effects. Twenty recommendations were made to address these issues.

Government recommendations on the Sydney third runway project (contained in the assessment report on the EIS)

- the establishment of a steering committee and preparation of a noise management plan
- long-term studies into the health risks associated with noise from the proposed runway
- management of construction times and monitoring for noise emissions
- preparation of an environmental management plan to manage impacts on Botany Bay—to be regularly reviewed and publicly advertised
- compliance with state pollution-turbidity standards
- consultation with New South Wales Maritime Services Board about designs for dredging operations
- clear outline of sea-grass recolonisation processes and goals
- outline of the feasibility of alternative habitats for migratory waders
- extension of the program for bird monitoring and management
- improvement of emergency plans for accidents
- establishment, prior to construction, of air-quality monitoring stations
- establishment of a dust monitoring network
- further studies into baseline water-quality data, benthic communities in Botany Bay, remedial measures to reduce wave impacts on Lay Robinsons Beach, identification of alternative relocation sites for regionally rare saltmarsh species, developing means to record ambient odours

The assessment of the proposal concluded that:

> The environmental impacts of the proposal are . . . the subject of controversy and the issues cannot in all instances be resolved definitively. Further studies will be necessary into aspects of the construction and operation of the third runway should there be a decision to proceed with the proposal. Taking all these matters into account, the Department has concluded that the environmental impacts of the proposal have been examined as fully as practicable and the object of the . . . Act has therefore been met.
>
> (DASETT 1991, p. 7).

The Minister for Shipping and Aviation Support was then obliged to take the recommendations into account in making a decision, and the proposal was finally approved by Cabinet, subject to adoption of the assessment report recommendations (Jenkins 1993). It was then the responsibility of the Federal Airports Commission to determine whether the proposal would proceed.

Following the approval and construction of the third runway there has been intense public controversy that has received significant attention in the national media. By the end of 1995, over 200 newspaper articles had been published on the issue, together with academic literature assessing the process and impacts. For instance, Johnson (1991) argues that the EIA process for the third runway was simply treated as a hurdle after the decision had already been made.

RECENT REVIEWS AND PROPOSED CHANGES TO COMMONWEALTH EIA LEGISLATION

In October 1993 the Commonwealth Minister for Environment announced a major review of the EIA process to facilitate ecologically sustainable development, improve environmental protection measures and to ensure consistency with the *Intergovernmental Agreement on the Environment* (see Chapter 8). The Commonwealth Environmental Protection Agency (CEPA) released an initial discussion paper defining the scope of the

Figure 7.4 Photograph of Sydney third runway
Source: supplied by Federal Airports Corporation

review, 'Setting the Direction', in November 1993. Public input was sought on the following matters:

- the objectives of environmental impact assessment
- the appropriate role for the Commonwealth government in environmental impact assessment
- issues that should be covered by the review
- the appropriate guiding principles that should govern the development and administration of the Commonwealth EIA process.

After receiving about 100 submissions on this paper, CEPA adopted eight guiding principles to direct any reform of the EIA process:

- to provide real opportunities for public participation in government decision-making
- to be open and transparent
- to provide certainty of application and process to all participants, including the community, governments, industry, and project proponents
- to provide accountable decision-making
- to be administered with integrity and professionalism
- to provide cost-effective processes and outcomes
- to be flexible enough to deal effectively and efficiently with all proposals assessed
- to ensure practical outcomes for effective environmental protection (CEPA 1994, p. iii)

At the same time as CEPA released its initial discussion paper, it also arranged six consultancies to examine particular issues for the review. The titles of the papers produced by these consultancies are as follows:

- 'Public Participation in the Environmental Impact Assessment Process: Public Inquiry Process'
- 'Social Impact Assessment'
- 'Assessment of Cumulative Impacts and Strategic Assessment in Environmental Impact Assessment'
- 'An Analysis of EIA Practices and Procedures in Australian States and Territories'
- 'Analysis of Environmental Impact Assessment Procedures in Other Countries'.

Options for reform were released in a subsequent discussion paper published in November 1994. This paper is divided into three parts: Part I provides the background and context of the review; Part II, which comprises the bulk of the paper, identifies three levels at which changes could be made and identifies a range of options at each level; and

Part III provides a very brief (one page) comment on future directions for EIA in Australia (CEPA 1994). At the same time as the discussion paper was being circulated for public comment, a series of 11 public information seminars were held in Australia's capital cities (a total of 506 people attended) to explain the review process, to let people know how to make a submission, and to give them an opportunity to raise issues and ask questions. A summary report from these seminars was prepared outlining the issues raised (Change Agency 1995).

The main discussion paper (CEPA 1994) includes a number of reform options that would expand the role of the Commonwealth EIA process. For example, one option is to give the Commonwealth Minister for Environment the discretionary power to require assessment of any project likely to raise environmentally significant issues of national or international importance. There is also a proposal to change the triggering mechanism by the introduction of a schedule of designated Commonwealth actions or decisions likely to result in significant environmental impacts. In addition, the Environment Minister, in consultation or agreement with the action Minister, would have the power to determine that a proposal is likely to be environmentally significant, even if it was not on the designated lists.

A number of procedural reforms were also proposed, such as:

- enabling the Environment Minister to set legally binding conditions in consultation with the action Minister
- setting a time period of 20 working days (after receipt of a notice of intention) within which a determination on the need for assessment must be made
- establishment of more stringent monitoring requirements
- screening criteria for assessment of cumulative impacts
- provision for audits of the environmental conditions set by the Minister
- removal of standing requirements in relation to judicial review of decisions sought by any person.

In all there are 19 procedural reform options listed in the main discussion paper (CEPA 1994, pp. 50–1). There are also eight reform options relating to EIA objectives, jurisdiction of Commonwealth legislation, and triggering mechanisms.

Following the release of the discussion paper, a number of public submissions were received. A reform package has been developed based on the discussion paper, the public responses, and internal discussions within CEPA and DEST. The reform package was not publicly available and the momentum for reform was perhaps lost among the broader investigations of the Commonwealth's roles and responsibilities in relation to both states and territories, and to environmental matters in general.

8

Prospects for Australian EIA in the twenty-first century

Overview of the differences between jurisdictions

The foregoing chapters demonstrate that EIA legislation and practice varies considerably between jurisdictions. Consideration of the prospects for EIA in Australia requires a broad grasp of these variations. Therefore, before the discussion of the future of EIA in Australia, the main areas of variation are recapped below.

Triggering

The triggering process is discretionary for all EIA processes in Australia, with the exception of New South Wales (Part IV of the New South Wales Environmental Planning and Assessment Act has a prescriptive list of designated developments requiring assessment), and the Australian Capital Territory, where 'defined decisions' automatically require a preliminary assessment. In other cases, the decision to trigger the Act is often the responsibility of the responsible authorities, with the final decision to require an assessment resting with either an Environment/Planning Minister or a Governor.

In Queensland, however, the decision to require an EIA may be made by either local government or the responsible authority (depending on the Act). In New South Wales, under Part V of the Environmental Planning and Assessment Act, the requirement may be made by the relevant Minister.

Both Western Australia and Victoria have broad referral powers, whereby any person may refer a proposal for consideration. The Northern Territory, Western Australia, and New South Wales (Part IV of the Environmental Planning and Assessment Act) also have powers for the Environment/Planning Minister to require EIA.

Scoping

In New South Wales, Queensland, Tasmania, and the Commonwealth, scoping of issues is not required by legislation, although it sometimes occurs in practice. Australian Capital Territory legislation provides for public exhibition, but there is no provision for public comments and revision of guidelines. Public scoping is available in the Northern Territory, South Australia, Tasmania (for level 3 assessments), Victoria, and Western Australia.

Levels of assessment

The levels of assessment also vary, with only Queensland having one assessment level (the EIS equivalent) for its two Acts. The Australian Capital Territory has the greatest number of formal assessment levels (PA, PER, EIS, and inquiry). Western Australia has a multi-graded system using different terminology relative to other EIA systems (CER, PER, and ERMP), together with inquiries and informal advice.

Terminology

The terminology used to describe the processes involved also differs between jurisdictions: Victoria has EESs; Queensland IASs; South Australia DRs; and Tasmania has development proposal and environment management plans, SECISs, and IARs. The most common levels are the PER (five jurisdictions), the EIS (seven jurisdictions), and the inquiry (six jurisdictions). In practice, no inquiries have been called by any of the states and territories. This is significantly different to the Commonwealth level, where five have been completed. There are also differences in the numbers of

EIA documents at different levels, with New South Wales completing more EISs than all the other states combined.

Quality of documents and public review

In practice there is variation in the quality of Australian EIA documentation, although little research has been done on either EIA documents or the quality control of EIA consultants. Once the documents have been prepared, public participation and consultation usually occurs: most EIS-equivalent documents are publicly exhibited for approximately one month. Exceptions to this one-month public review period include Western Australia with ten weeks, and South Australia with six weeks. It is interesting to note that the lower PER level in Western Australia requires a longer period (eight weeks) than the higher EIS levels in South Australia (six weeks), New South Wales (30 days), the Northern Territory (28 days), and the Commonwealth (28 days). Public exhibition is discretionary in Queensland under the State Development and Public Works Organisation Act and in Victoria (although two months is the usual practice in Victoria). New South Wales and Queensland, and assessment levels 1 and 2 in Tasmania, do not have formal requirements for the proponent to respond to public submissions. In other cases, the requirement that the proponent respond to public submissions only applies to the EIS (not to the PER or equivalent). Western Australia is relatively unique in that the proponent is required to respond to submissions made in relation to all assessment levels.

Weight given to the EIA

In some cases assessment is advisory only: recommendations are only taken into account by an internal decision maker (for example, in the Northern Territory and in the Commonwealth jurisdiction). In Western Australia, however, an 'independent' decision is made by the Environment Minister, following consultation with other relevant Western Australian ministers. Similarly, environmental recommendations in the Australian Capital Territory are tabled in the Legislative Assembly for determination. In South Australia an 'independent' decision is made by the Governor, while in Tasmania, the final determination is made by either the Board of Environmental Management (for level 2 assessments) or the Governor (level 3 assessments). Here it is interesting to note that Tasmania is unique in providing for public comment (28 days) on a draft assessment report (level 3) before a final assessment report may be produced. New South Wales is also different in this respect: in the event that the proponent is also the determining authority, Part V of the Environmental Planning and Assessment Act requires a combination of independent external decision-making and internal decision-making. Although the determining authority makes the final decision, the proposal cannot proceed in the event that the Director General of the Department of Urban Affairs and Planning refuses the project.

Monitoring and appeals

Formal monitoring requirements appear to be weak in some states, with no formal legislative provisions for monitoring in Queensland or Victoria, although monitoring does occur in practice. The Australian Capital Territory has indirect monitoring requirements. In contrast, Western Australia, the Commonwealth, South Australia, and the Northern Territory have explicit monitoring requirements in their Acts.

Appeal rights also vary considerably. While none are available in Tasmania (in relation to level 3 assessment), South Australia, or Queensland(under the State Development and Public Works Organisation Act), limited provisions are evident in the Northern Territory and Tasmania (levels 1 and 2). Appeal rights are also available in Queensland under the Local Government (Planning and Environment) Act. Appeal rights for the Commonwealth process are unclear given the debate about the legal status of the Administrative Procedures. New South Wales is unique in having broad standing rights for appeal.

CHANGES TO EIA LEGISLATION

EIA legislation in Tasmania, Queensland, Western Australia, and the Commonwealth is currently undergoing major changes, although changes in Queensland and the Commonwealth jurisdiction are on hold for political reasons.

In Tasmania a new Act, the Environmental Management and Pollution Control Act 1994, has replaced the Environment Protection Act 1973. In Queensland a draft Bill, the Planning, Environment and Development Assessment Bill, is currently being considered, and if it is adopted it will repeal the Local Government Planning and Environment Act. The process of EIA under the Queensland State Development and Public Works Organisation Act is also currently being amended: the

Administrative Procedures are being redrafted, outlining more specific procedures (Planning and Assessment Section, pers. comm., May 1997). However, these Queensland amendments are now on the 'back-burner' due to someinconsistencies with the Planning, Environment and Development Assessment Bill. In Western Australia, the EIA process was amended through the Planning Legislation Amendment Act 1996 and associated amendments to the Environmental Protection Act. These effectively gave the EPA power to assess statutory town-planning schemes and amendments in a parallel process to the planning approval process.

Finally, the Commonwealth may lose the impetus for reform after a major review of EIA initiated by the Keating government was re-examined by the Howard government (also re-examined were issues of Commonwealth roles and responsibilities in environmental matters). The Commonwealth EIA changes, originally scheduled for the start of 1996, are likely to be downgraded in priority. The priorities of the Howard government mean that the Commonwealth EIA changes, originally scheduled for the start of 1996, are likely to be downgraded in priority. They are unlikely to be in place by the twenty-first century.

THE CASE-BY-CASE NATURE OF EIA

The case studies discussed in Chapter 6 demonstrate that the operation of EIA in practice can be quite complex, once developers, local government, consultants, the public, state governments, politicians, and other stakeholders become involved. While many projects will be dealt with routinely by the EIA process, each project has its own unique set of environmental impacts dependant on the locational setting, the scale of the project (in time and space) and the nature of impacts. In the case of Magnetic Quay, for example, the complications of multiple jurisdictions were exacerbated by inadequate investigations as well as by the need for a number of studies and different levels of EIA (none of which were able to resolve the negative environmental impact resulting from the project). The case of Hindmarsh Island demonstrates how significant public involvement in a project can occur long after the formal public involvement in the EIA process has been completed and the project approved. In some cases (Tindal Airbase, Magnetic Quay) the overlapping roles of Commonwealth and state EIA legislation needed to be resolved. In the case of the Sydney third runway, this overlap in legislation was even subject to a legal challenge.

AUSTRALIA'S NATIONAL APPROACH TO EIA

During the 1990s, largely in response to issues of overlapping EIA jurisdiction, there have been a number of national initiatives in relation to EIA. It had long been clear that there was a need for a more coordinated approach to EIA, but because of the complexity of the different jurisdictions and the variations in EIA practice, there was uncertainty as to how this coordination could be delivered. The 1990s were very important for Australian EIA for a number of reasons.

First, at the start of the decade, there was a recognition by all Australian governments that there was a need for cooperation on environmental matters. Second, the 1992 'World Summit' in Rio provided an impetus for the Australian government to demonstrate its environmental credentials in a number of contexts, including EIA. Third, Australian EIA practitioners were being increasingly drawn into a broader global EIA network, particularly with contributions to capacity building and international research into the effectiveness of EIA. For these reasons, the 1990s have produced a number of important EIA documents, which are discussed below.

The *National Approach on EIA*

The *National Approach on EIA*, was produced in 1991. The impetus for this came from a special premiers' conference held in 1990. At this conference it was established that there was a need for an intergovernmental agreement on the environment that would include the development of a national approach to environmental impact procedures throughout Australia and greater streamlining of EIA and other approval processes. In response to this conference, the Australia and New Zealand Environment and Conservation Council (ANZECC) established a working group to prepare this national approach. The working group comprised EIA practitioners from individual states, territories, and the Commonwealth. Taking note of concerns, expressed at the conference, that the adoption of a national approach would be too prescriptive (Bailey and English 1991), it took a pragmatic approach.

The working group prepared a background paper (ANZECC 1991a) outlining the objectives of EIA, its terminology, and time-frames for the development of the national approach. The background paper also included a discussion of national issues associated with the EIA process. The working group met with a group of selected proponents and held various consultations, mostly through individual members of the working group consulting key organisations and persons at the state and territory level. A sub-group of the working group also met with national bodies and Commonwealth agencies. A draft of the national approach document was then circulated for comment and a number of issues raised from this process were taken into account in the preparation of the final document.

Another EIA working group (one of 37 specialist groups) was set up early in 1992 under the umbrella of the intergovernmental ecologically sustainable development (ESD) committees (discussed in detail below). However, the main role of that group was to provide input to the ESD implementation program and to contribute to the ESD and Greenhouse strategies rather than to produce another public document on EIA.

Sections 1–2 of the *National Approach on EIA* (ANZECC 1991b) provide an introduction, a description of terminology used, and an outline of the objectives of EIA in Australia. The objectives are:

- to ensure that decisions are taken following timely and sound environmental advice
- to encourage and provide opportunities for public participation in environmental aspects of proposals before decisions are taken
- to ensure that proponents of proposals take primary responsibility for protection of the environment related to their proposals
- to facilitate environmentally sound proposals by minimising adverse impacts and maximising benefits to the environment
- to provide a basis for ongoing environmental management, including through the results of monitoring
- to promote awareness and education in environmental values (ANZECC 1991b, pp. 3–4).

The last objective relating to environmental awareness and education does not appear to have been given such prominence in later agreements.

Section 4 of the *National Approach on EIA* provides a very brief comment on ESD (discussed below), and section 5 discusses national principles for EIA in Australia. These principles relate to assessing authorities, proponents, the public, and government.

National principles for EIA in Australia

Principles for assessing authorities

- Provide clear guidance on types of proposals likely to attract environmental impact assessment and on levels of assessment.

- Provide proposal-specific guidelines (or a procedure for their generation) focused on key issues, and incorporating public concerns and a clear outline of the IEA process—amendments to guidelines should only be based on significant issues arising after guidelines have been adopted.
- Provide guidance, to all participants in the EIA process, on criteria for environmental acceptability of potential impacts, including such things as the principles of ecologically sustainable development, maintenance of environmental health, relevant local and national standards and guidelines, and codes of practice and regulations.
- Negotiate with key participants to set an assessment timetable on a proposal-specific basis, and commit to using best endeavours to meet it.
- Seek and promote public participation throughout the process, with techniques and mechanisms tailored appropriately to specific proposals and specific publics.
- Ensure that the total and cumulative effects of using or altering community environmental assets (for example air, water, amenity) receive explicit consideration.
- Report publicly on the assessment of proposals.
- Ensure that predicted environmental impacts are monitored, the results assessed by a nominated responsible authority, and feedback is provided to improve continuing environmental management of proposals.
- Monitor properly the efficiency and effectiveness of the environmental impact assessment process, so as to learn from the past, to streamline requirements and to help maintain consistency.
- Review, adapt and implement techniques and mechanisms that can improve the process and minimise uncertainty and delays.
- Ensure that educational opportunities inherent in the EIA process are actively pursued.

Principles for proponents

- Take responsibility for preparing the case for assessment of the proposal.
- Consult the assessing authority and the community as early as possible.
- Incorporate environmental factors fully into proposal planning, including a proper examination of reasonable alternatives.

- Agree on a proposal-specific evaluation timetable and commit to using best endeavours to meet it.
- Take the opportunity, offered by the EIA process, to improve the proposal environmentally.
- Make commitments to avoid (where possible) and otherwise minimise, ameliorate, monitor, and manage environmental impacts—and implement these commitments.
- Amend environmental management practices responsibly, following provision and dissemination of environmental monitoring results.
- Identify and implement responsible corporate environmental policies, strategies, and management practices, with periodic review.

Principles for the public

- Participate in the evaluation of proposals through offering advice, expressing opinions, providing local knowledge, proposing alternatives, and commenting on how a proposal might be changed to better protect the environment.
- Become involved in the early stage of the process, as that is the most effective and efficient time to raise concerns; participate in associated (and earlier) policy, planning, and program activities as appropriate, since these influence the development and evaluation of proposals.
- Become informed and involved in the administration and outcomes of the environmental impact assessment process, including being involved in the following: assessment reports of the assessing authority, policies determined, approvals given and conditions set, monitoring and compliance audit-activities, environmental advice, and reasons for acceptance or rejection by decision makers.
- Take a responsible approach to opportunities for public participation in the EIA process, including the seeking out of objective information about issues of concern.

Principles for government

- Provide policy and planning frameworks that set contexts for the environmental assessment of proposals.
- Base decisions on proposals with potentially significant environmental impact on advice resulting from the EIA process, and

include provisions for effective protection and management of the environment.

- Apply the EIA process equally to proposals from both the public and private sectors.
- Within each jurisdiction (Commonwealth, state/territory) provide for a coordinated government decision-making process to which the outcomes of EIA can be directed—and develop mechanisms to synchronise processes for decision-making such that, in relation to proposals requiring multiple approvals, where possible, the opportunity exists for decisions to be made in parallel rather than sequentially.
- Ensure assessment reports are available to the public before or at the time of decision-making.
- Establish one national agreement to ensure a single orderly process is in place where the EIA responsibilities of several governments are involved.
- Provide support, if and when appropriate, to participants in the process to enable better and informed involvement.
- Provide opportunities for reasonable public and proponent objections, on decisions made other than at Ministerial level, regarding the requirement for and level of assessment, adherence to due process, and environmental advice given to decision makers
- Implement this national approach including, where appropriate, progressive amendment of statutory provisions to increase consistency in the process.
- Maintain the integrity of the EIA process

ANZECC 1991a, pp. 5–8

Section 5 also contains a discussion of a number of areas where Australian legislation has been deficient, such as cumulative effects, monitoring, compliance-audit activities, coordinated government decision-making incorporating EIA outcomes, and project-specific evaluation timetables.

Section 6 deals with the proposals for national practice of EIA in Australia, based on the preceding discussion of principles. The fact that it suggests incorporating EIA principles into policies and major programs is consistent with the definition of EIA given at the start of this book. Other authors (for example, Therivel et al. 1992) refer to this practice as strategic environmental assessment (SEA). It currently has limited practice in Australia (Wood 1992). Section 6 suggests that:

- there needs to be clearer criteria for initiating the EIA process
- the Minister or agency responsible for the process should have the power to initiate it
- there should be different levels of publicly accountable EIA
- a time-schedule for the process should be set out in advance.

This section also discusses outcomes, proposal facilitation, and specifically addresses (section 6.7) the need for a national agreement between the states, territories, and the Commonwealth. Finally, section 7 of the *National Approach* proposes that the document should be reviewed by ANZECC within five years of its adoption.

Linkages with the Commonwealth government's strategy for ecologically sustainable development

Sustainable development has been defined by the World Commission on Environment and Development, in the document *Our Common Future*, as development that 'meets the needs of the present without compromising the ability of future generations to meet their own needs' (WCED 1987, p. 8). In order to achieve critical sustainable development objectives for environment and development policies, it is important to use strategies such as conserving and enhancing the resource base, reorienting technology, managing risk, and merging environment and economics in decision-making (WCED 1987, p. 93). The last of these strategies is particularly relevant for EIA, as it promises an integrative process that can help avoid environment and development problems arising from the sectoral fragmentation of responsibility.

In a discussion paper entitled *Ecologically Sustainable Development*, the Commonwealth of Australia coins its own term, 'ecologically sustainable development' (ESD) (Commonwealth of Australia 1990), and defines the broad areas of operational principles that it considers necessary for the implementation of a sustainable society. These operational principles are:

- improvement in material and non-material well being
- inter-generational equity
- intra-generational equity
- maintenance of ecological systems and protection of biodiversity
- global ramifications, including international spill-overs, international trade, and international cooperation
- dealing cautiously with risk, uncertainty, and irreversibility.

In 1990 the Commonwealth government established nine working groups to consider the implementation of ESD principles in sectors of Australia's economy that have major impacts on the environment. This sectoral approach was strongly criticised by Australian conservation groups (Hare 1990). The nine working groups released their final reports in December 1991. The ESD reports, because of their sectoral nature, are variable in their analysis of linkages with the EIA process, but a few of them contain detailed recommendations for EIA. It is not apparent from the final ESD report (Commonwealth of Australia 1990) that any strong linkages between EIA and ESD have been identified for further development.

The linkages between this ESD strategy and the EIA process are not made clear in the *National Approach on EIA* (ANZECC 1991a), which gives only passing mention to ESD in the set of principles relating to assessing authorities. The *National Approach* suggests that EIA is just one of many methods for achieving ESD. It also outlines some of the major connecting points where EIA can assist in achieving ESD.

Connections between EIA and ESD

The use of resources by present generations is achieved while protecting the interests of future generations through, for example:

- maintaining and enhancing natural capital (for example clean water, clean air, uncontaminated soil)
- avoiding over-exploitation of renewable resources
- minimising waste
- protecting biodiversity and ecosystem integrity
- providing net community benefits from proposals that are implemented
- social equity—for example, through public participation in the decision-making process
- reflection of full environmental costs of proposals in decisions on resource use
- caution in dealing with environmental risk and irreversibility.

ANZECC 1991a, p. 4.

The background paper to the *National Approach on EIA* attempts to illustrate, in diagrammatic form, EIA within the context of an ESD framework (ANZECC 1991a, p. 4), to show how sustained outcomes are achieved. The background paper also lists some of the key issues in the relationship between EIA and ESD. These issues, which were raised mostly in response to a draft of the national approach, are:

- confusion over the definition of 'sustainable'
- the need for a consistent approach to both EIA and ESD
- the need for a precautionary principle
- the need to avoid the situation of economic factors overriding environmental ones
- the need for EIA to have more comprehensive parameters if it is to be part of ESD
- the need for EIA to include many principles and strategies of ESD (ANZECC 1991b, p. 8).

However, not all of these suggestions were incorporated into the *National Approach on EIA* (ANZECC 1991a).

While all nine sectoral ESD reports were intrinsically concerned with the environmental impact of the relevant sector of the economy, the importance of EIA as a process within each sector is interpreted quite differently. For this reason, not all of the reports make recommendations about linkages between the relevant sector of the economy and EIA, although most make reference to the EIA process. From an examination of the nine ESD reports, Harvey (1992b) concludes that it is apparent that the EIA process has been interpreted by the various working groups as being more important for some sectors of the economy, such as tourism and mining, than it is for others, such as agriculture, energy use, and fishing. Of those ESD reports that specifically address the EIA process, Harvey (1992b) identifies the following common threads:

- a need for a more streamlined, integrated, and consistent approach to EIA between the states and the Commonwealth—this includes the development of credible, clear, and open criteria
- a need for greater certainty about the triggering mechanism for EIA and clearer guidelines on the level of assessment required
- a need for EIA to consider cumulative and long-term impacts
- a need for public participation early in the process in order to be effective
- a need for some kind of post-development auditing of projects which have undergone the EIS process (Harvey 1992b, p. 270).

It is evident that each of the above five principles had already been identified to some extent in the *National Approach to EIA* in the section dealing with national principles for assessing authorities, for proponents, for the public, and for government. The *National Approach* does not, however, specifically address these principles within the context of ESD. It does not offer a new EIA process, but it does try to outline a consensus, on principles of EIA, between the commonwealth, the states and the territories. The identification of linkages between EIA and ESD in the *National Approach* appears to be a reactive approach to the perceived need to identify those linkages rather than an innovative approach that takes the opportunity to propose stronger linkages.

It is apparent that there has been an attempt to identify linkages between EIA and ESD in both the *National Approach on EIA* and the nine ESD reports. While these documents are welcome initiatives, there is a danger that linkages between EIA and ESD may not be developed far enough. The *National Approach* identifies some of the linkages but does not take this further to suggest that the proposed national agreement on EIA should include ESD principles. The *National Approach* makes no recommendation for this proposed agreement to include factors such as:

- integration of ecological and economic factors
- trans-boundary environmental impacts
- cumulative environmental impacts
- auditing of environmental impacts.

These are essential for the implementation of ESD through the EIA process. Although EIA is only a part of ESD implementation, stronger linkages would help in extending the application of EIA to policies.

The Australian *Intergovernmental Agreement on the Environment*, 1992

In May 1992 state and territory ministers endorsed an *Intergovernmental Agreement on the Environment* to establish rules for the interaction of Commonwealth, state, territory, and local governments on environmental matters. The agreement is divided into four sections:

1. application and interpretation
2. roles of the parties
3. responsibilities and interests
4. principles of environmental policy, and implementation and application of these principles.

In addition there are nine schedules, all of which are to be interpreted in accordance with sections 1–3 of the agreement. All parties signed the agreement, with the Northern Territory reserving agreement in relation to sections on vehicle emission control. The schedules relate to:

- data collection and handling
- resource assessment
- land-use decisions
- land-use approval processes
- environmental impact assessment
- national environment protection measures
- climate change
- biological diversity
- national estate (a list of sites of particular historic, scientific, or social significance)
- world heritage
- nature conservation.

Schedule 3, which deals with EIA, consists of five parts. Parts 1 and 2 relate to the scope of EIA, the need for consistency, and the need to avoid duplication in the process:

> The parties agree that it is desirable to establish certainty about the application, procedures and function of the environmental impact assessment process, to improve the consistency of the approach applied by all levels of Government, to avoid duplication of process where more than one Government is involved and interested in the subject matter of an assessment, and to avoid delays in the process . . . The parties agree that impact assessment in relation to a project, program or policy should include, where appropriate, assessment of environmental, cultural, economic, social and health factors.
>
> IGAE 1992, p. 21

Part 3 of Schedule 3 relates to consistency in the application of EIA processes across different jurisdictions.

Schedule 3, Part 3, of the Intergovernmental Agreement on the Environment (1992)

The parties agree that all levels of Government will ensure that their environmental impact assessment processes are based on the following:

i. the environmental impact assessment process will be applied to proposals from both the public and private sectors;
ii. assessing authorities will provide information to give clear guidance on the types of proposals likely to attract environmental impact assessment and on the level of assessment required

iii. assessing authorities will provide all participants in the process with guidance on the criteria for environmental acceptability of potential impacts including the concept of ecologically sustainable development, maintenance of human health, relevant local and national standards and guidelines, protocols, codes of practice and regulations

iv. assessing authorities will provide proposal specific guidelines, or a procedure for their generation, focussed on key issues and incorporating public concern together with a clear outline of the process

v. following the establishment of specific assessment guidelines, any amendments to those guidelines will be based only on the significant issues that have arisen following the adoption of those guidelines

vi. time schedules for all stages of the assessment process will be set early on a proposal specific basis, in consultations between the assessing authorities and the proponent

vii. levels of assessment will be appropriate to the degree of environmental significance and potential public interest

viii. proponents will take responsibility for preparing the case for assessment of a proposal and for elaborating environmental issues which must be taken into account in decisions, and for protection of the environment

ix. there will be full public disclosure of all information related to a proposal and its environmental impacts, except where there are legitimate reasons for confidentiality including national security issues

x. opportunities will be provided for appropriate and adequate public consultation on environmental aspects of proposals before the assessment process is complete

xi. mechanisms will be developed to seek to resolve conflicts and disputes over issues which arise for consideration during the course of the assessment process

xii. the environmental impact assessment process will provide a basis for setting environmental conditions, and establishing environmental monitoring and management programs (including arrangements for review), and developing industry guidelines for application in specific cases

Part 4 of Schedule 3 provides the under-pinnings for the development of a national agreement on EIA (discussed below):

> A general framework agreement between the Commonwealth and the States on the administration of the environmental impact assessment process will be negotiated to avoid duplication and to ensure that proposals affecting more than one of them are assessed in accordance with agreed arrangements.
>
> (IGAE 1992, p. 22).

Part 5 of Schedule 3 refers to accreditation of EIA processes between the Commonwealth and the states.

Schedule 4 of the *Intergovernmental Agreement* provides for the setting up of a National Environment Protection Authority. As provided for in Schedule 4, The National Environment Protection Authority, which came into being as the National Environment Protection Council in August 1995, is a Ministerial Council with one Minister from each of the States, Territories, and the Commonwealth. It is responsible for establishing nationally applied measures for environmental protection matters and for producing agreed national environmental quality standards and monitoring their enforcement and implementation. It also takes an active part in environmental research, environmental education, and establishing common data bases, as well as taking responsibility for the state of environment monitoring.

Guidelines and Criteria for Determining the Need for and Level of EIA in Australia, 1996

The aim of the *Guidelines and Criteria* (ANZECC 1996) is greater consistency in applying the EIA process in Australia. It provides a framework for determining 'whether and at what level formal EIA procedures will apply to potentially significant proposals'. This document provides a common basis for screening projects and actions, and for deciding on the level of assessment required.

The document outlines seven criteria intended to be used primarily by practitioners of EIA working in assessing authorities. They are also intended to guide situations where the decisions being made are discretionary.

Criteria for environmental impact assessment, *Guidelines and Criteria for Determining the Need for and Level of EIA in Australia*

- The character of the receiving environment

- The potential impacts of the proposal
- Resilience of the environment to cope with change
- Confidence of prediction of impacts
- Presence of planning or policy framework or other procedures providing mechanisms for managing potential environmental impacts
- Other statutory decision-making processes that may provide a forum to address the relevant issues of concern
- Degree of public interest

ANZECC 1996

The criteria can be used for the following:

- decisions as to whether or not to bring the proposal to the attention of the assessing authority
- decisions by the assessing authority as to whether or not to recommend, or require, that the proposal should be subject to formal EIA procedures
- recommendations by the assessing authority in relation to the level of EIA to be applied
- determinations by the assessing authority as to the level of EIA to be applied.

In cases where there is very little discretion, such as where the relevant legislation contains specific lists of criteria or categories of proposals subject to various levels of EIA, the broader criteria can be used to refine the specific lists in order to achieve national consistency.

The *Guidelines and Criteria* is structured around a series of questions for consideration together with a checklist of issues to be considered (see figure 8.1). While the questions are a useful guide to determining the importance of each of the seven criteria, there is a danger that the checklists of issues could be seen as definitive lists rather than as a guide. This danger is particularly evident in relation to the tendency to treat environmental matters in a sectoral manner. Take, for example, question 2—'is it an existing or potential environmentally significant area?'. This question must be considered under criteria 1—'character of the receiving environments (natural and human)'—which goes on to split up the biophysical environment into geomorphological characteristics and ecological systems. Under 'geomorphic characteristics' come wetlands, lakes, coastlines and dunes, islands, rivers or estuaries, plateaus, alpine areas, desert area, and

Figure 8.1 Detailed questions relating to criteria for EIA
Source: ANZECC 1996

CRITERIA FOR THE DETERMINATION OF THE NEED FOR AND LEVEL OF ENVIRONMENTAL IMPACT ASSESSMENT IN AUSTRALIA

1. CHARACTER OF RECEIVING ENVIRONMENT	2. POTENTIAL IMPACTS PROPOSAL	3. RESILIENCE OF NATURAL AND HUMAN ENVIRONMENTS TO COPE WITH CHANGE	4. CONFIDENCE OF PREDICTION OF IMPACTS	5. PRESENCE OF PLANNING, POLICY FRAMEWORK AND 6. OTHER STATUTORY DECISION MAKING PROCESSES	7. DEGREE OF PUBLIC INTEREST
Consider: Is it, or is it likely to be, part of the conservation estate or subject to treaty? Is it an existing or potential environmentally significant area? Is it vulnerable to major natural or induced hazards. Is it a special purpose area? Is it an area where human communities are vulnerable? Does it involve a renewable or a non renewable resource? Is it a degraded area, subject to significant risk levels, or a potentially contaminated site?	**Consider:** Will construction, operation and/or decommissioning of the proposal have the potential to cause significant changes to the receiving environment? (on site or off site, short term or long term). Could implementation of the proposal give rise to health impacts or unsafe conditions? Will the project significantly divert resources to the detriment of other natural and human communities?	**Consider:** Can the receiving environment absorb the level of impact predicted without suffering irreversible change? Can land uses at and around the site be sustained? Can sustainable uses of the site be achieved beyond the project life? Are contingency or emergency plans proposed or in place to deal with accidental events?	**Consider:** What level of knowledge do we have on the resilience of a given significant ecosystem? Is the project design and technology sufficiently detailed and understood to enable the impacts to be established? Is the level and nature of change on the natural human environment sufficiently understood to allow the impact of the project to be predicted and managed? Is it practicable to monitor predicted effects? Are present community values on land use and resource use likely to change?	**Consider:** Is the proposal consistent with existing zoning of the long term policy framework for the area? Do other statutory approval processes exist to adequately assess and manage project impacts. What legislation, standard codes or guidelines are available to properly monitor and control operations on site and the type or quantity of the impacts?	**Consider:** Is the proposal controversial or could it lead to controversy or concern in the Community? Will the amenity, values or lifestyle of the community be adversely affected? Will the proposal result in inequities between sectors of the community?
NOTE: Offsite as well as onsite characteristics should be considered, where relevant.	**NOTE:** This should include consideration of the magnitude of the impacts, their spatial extent, the duration and intensity of change, the total product life cycle and whether and how the impacts are manageable.	**NOTE:** Cumulative as well as individual impacts should be considered in the context of sustainability.			

karst. It may be difficult to know what to do with remnant landscape features (such as glacial erratics or granitic tors) or geological type sections exposed in a cliff. Still in criteria 1, 'ecological systems' includes 'wilderness' as a category. Wilderness is, however, a purely human construct.

It is important to note that the criteria are not intended to specify thresholds in determining the need for formal EIA procedures. It is intended that the evaluation of individual proposals should be undertaken jointly with the assessing authority, the proponent, and other interested parties. A major advantage of the criteria and the associated questions and lists is that it makes the process more transparent. The assessing authority, the proponent, and the public can all use the criteria as a method of accountability in determining the need for, and the level of, EIA. The proponent will be aware of the issues that will be raised, but at the same time the assessing authority will need to be thorough in its evaluation of the criteria. Although evaluation of the criteria is subjective, it is possible that increased use of the criteria will lead to legal battles over their application. If this happens it is likely that experts will be used in court cases, and precedents regarding the application of the criteria will be set, thereby making the criteria more prescriptive.

The Australian *National Agreement on EIA*, 1997

In 1992, in accordance with the *Intergovernmental Agreement on the Environment* discussed above (see pp. 196–99), and in an attempt to produce consistency and avoid duplication, it was agreed to develop a national framework for EIA. The resultant *National Agreement on EIA* was finally completed and endorsed by ANZECC in June 1997. The aim of the agreement is to have a clear framework for coordinating assessments of proposals subject to EIA procedures that involve more than one government. The *National Agreement* has been delayed a few times over the five years of its evolution because of changing governments and their differing attitudes towards such an agreement. The *National Agreement* is divided into 15 parts.

The National Agreement on Environmental Impact Assessment (list of parts)

1. Parties to the agreement

2. Legal rights and obligations of the parties
3. Interpretation
4. Objectives of the agreement
5. Interests of parties
6. Cooperative arrangements
7. Accommodation of interests (Commonwealth–state)
8. Accommodation of interests (state–state)
9. Assessment procedures
10. Territorial seas
11. Schedules to the agreement
12. Freedom of information legislation
13. Application of the agreement
14. General consultation and cooperation
15. Further development and review

ANZECC 1997

The parties to the *National Agreement* (Part 1) are the state and territory governments and the Commonwealth government, but the agreement does not affect their legal or constitutional rights, and no legislative changes for EIA are introduced in any of the jurisdictions (Part 2). The *National Agreement* is set in the framework of the *Intergovernmental Agreement on the Environment* (Part 4) and draws on the *National Approach to EIA in Australia*. It replaces a number of bilateral agreements on EIA with one broader national agreement.

The *National Agreement* (Part 5) recognises that EIA usually involves only one government but also takes into account the fact that a party may have a legitimate interest in a proposal even if the project is not within its own jurisdiction—for example, a project affecting water quality in the Murray River. Parts 6–9 of the agreement deal with cooperative arrangements (and coordination of the assessment process) between:

- Commonwealth and state jurisdictions
- different states/territories.

The *National Agreement* also sets out mechanisms for dealing with the assessment of offshore proposals (Part 10).

Part 13 of the *National Agreement* provides a mechanism for resolving differences arising over the application of the agreement. Parts 14–15 provide for ongoing cooperation between governments on matters relating to the practice and procedure of EIA, and for a review of the *National Agreement* within five years of the date that it comes into effect. In summary, the *National Agreement* is a sensible approach to dealing with the differences in EIA legislation and procedures across Australia.

INTERNATIONAL STUDIES ON THE EFFECTIVENESS OF EIA AS A PROCESS FOR BETTER DECISION-MAKING

Recent attention has turned to the need for assessment of the overall effectiveness of the EIA process after more than 25 years of its formal operation. Glasson, Therivel, and Chadwick (1994) discuss ways of improving project-EIA effectiveness, with particular reference to the EC system and the UK system. One method for improving EIA effectiveness is to establish a review program for monitoring performance parameters at each stage in the EIA process. A study conducted by Wood (1995) examines the monitoring of EIA systems and checks if there is any feedback (designed to improve the effectiveness of the particular system) from the monitoring. Based on his evaluation criteria for EIA system monitoring, he concludes that, out of the seven jurisdictions he investigated, only the United States, the Netherlands, and Canada had requirements for the monitoring of the EIA process as well as mechanisms for incorporating feedback and, if necessary, modifying the system. He noted that each of these jurisdictions has a single body with overall responsibility for EIA and a legal duty to review or oversee the system.

The 1993–96 international effectiveness study

Most recently, a comprehensive international study (Sadler 1996) on the effectiveness of environmental assessment was conducted by the Canadian Environmental Assessment Agency, in conjunction with the International Association for Impact Assessment (IAIA). The study, conducted from 1993 to 1996, focused on the following ten themes:

- guiding values and principles
- application of sustainability concepts
- strategic environmental assessment
- cumulative and large-scale effects
- relationship to decision-making
- integrated approaches to impact analysis
- public participation and dispute settlement

- follow-up and post-project analysis
- total process management
- capacity building, with particular reference to developing countries (Sadler 1996, p. 9).

It had the following objectives:

- to review current issues, emerging trends, and future directions of EIA
- to examine the contribution of EIA to decision-making
- to document what works well with existing approaches
- to recommend cost-effective measures for improving EIA, with specific reference to the challenge of sustainable development.

The research design for the study was in four stages:

- a policy analysis of leading trends and issues
- an examination of the contribution of EIA to development decision-making
- an examination of operational excellence in the application of EIA methods, procedures, and components
- conclusions and guidelines for sound practice (Sadler 1996).

Methods used to collect information

A number of different methods were used to collect information for the study (Sadler 1996). Over the three years of the study, a number of conferences and workshops were held, including two in Australia. A major summit was held during the 1994 IAIA conference involving seven international organisations as well as the heads of EIA agencies from 26 countries.

A questionnaire survey on the status of EIA practice was sent to IAIA members and then to European, Australasian, and Francophone countries. In all there was a total of 324 responses, half of which were from IAIA members. Country and company 'status reports' from those attending the 1994 IAIA conference were consulted for the purpose of comparing EIA provisions, processes, and practice. Case studies and decision analyses were obtained from individual IAIA members, from partner agencies (EIA agencies affiliated with IAIA), and from developing countries (through the United Nations Environment Programme), Finally, reports and case studies on integrated EIA were collected from partner agencies.

The effectiveness study concludes that there are immediate, cost-effective opportunities to strengthen EIA in the following areas:

- establishing standards for quality performance in EIA

- upgrading EIA processes and activities to improve quality control
- public involvement and cumulative effects
- extending SEA as an integral part of decision-making
- sharpening SEA as a sustainability instrument (Sadler 1996).

ADOPTION OF FORMAL SEA IN AUSTRALIA OR A FOCUS ON BETTER PLANNING?

As noted above, the international effectiveness study, in which Australia was a key participant, suggests an increased focus on SEA as one method for strengthening the EIA process. Similarly, Therivel et al. (1992) and Glasson, Therivel, and Chadwick (1994) have discussed the need for SEA in the context of broadening the scope of EIA and overcoming some of the problems associated with project-based EIA. Therivel et al. (1992, p. 90) define SEA as

> the formalised, systematic and comprehensive process of evaluating the environmental effects of a policy, plan or programme and its alternatives, including the preparation of a written report on the findings of that evaluation, and using the findings in publicly accountable decision-making.

More recently, Therivel and Partidaro (1996) have reviewed the current status of SEA regulations and guidelines in 11 countries. They note that most SEAs have been carried out on a voluntary basis. Sometimes SEAs are supported by guidelines (for example, in the United Kingdom and Canada) but they are rarely required by regulation (for example, the United States and Netherlands). On a broader basis there are now requirements for SEA to accompany structural fund applications (applications for funds for broadly based projects relating to strategic or regional planning) through the European Union. Similarly, the World Bank requires SEA to be applied to regional and sectoral development activities (Therivel and Partidaro 1996).

In Australia there is potential for SEA at the Commonwealth level, although Wood (1994, 1995) and Therivel and Partidaro (1996) comment on the lack of it. McCarthy (1996) discusses the potential for SEA in South Australia, but the only examples of SEA in practice appear to come from Western Australia, New South Wales, and, perhaps, Victoria (Stone 1997; Krohn 1997). There is, however, an increasing move to implement SEA in Tasmania, in relation to assessments of planning schemes, and also in Queensland (Queensland Department of Environment, pers. comm., 1997)

The Australian ESD debate has involved recommendations that EIA in various sectors of the economy should be applied to policies, plans, and projects. In addition, the *National Approach on EIA* (ANZECC 1991a) suggests that a number of the principles and objectives outlined in the *Approach* could be applied to environmental assessment of policies and major programs. In fact, the Commonwealth Environment Protection (Impact of Proposals) Act encompasses 'the formulation of proposals . . . decisions and recommendations and . . . incurring of expenditure' (section 5(2)) and could clearly, therefore, be applied to policies and other actions. Similarly, the South Australian Development Act is applicable to a development or 'project' defined as 'an activity or circumstance that does not require approval under this Act' (for the purposes of the Act, 'project' is not synonymous with 'development'). This potential for SEA has generally not been adopted in Australia, even where there is a demonstrated need, such as in the energy sector (Harvey 1994b), perhaps because of a 'closed-shop' mentality in relation to political and bureaucratic policy formulation.

However, there are many examples of Australian approaches to strategic planning that incorporate SEA principles but do not have the SEA tag. These include the integrated approach to developing management strategies for the Great Barrier Reef Marine Park or the Murray Darling Basin (RAC 1993), both of which required detailed environmental assessment and involved the juxtaposition of Commonwealth and state jurisdictions. On a smaller scale, the development of a marina strategy in South Australia (Harvey & Swift 1990a; Harvey 1992a, 1993a) arose from a plethora of marina-development applications which were being dealt with by project-specific EIAs rather than a broader-based strategic environmental approach. Thus, the lack of formal SEA in Australia may reflect either the existence of such ad hoc SEAs or may be related to the use of strategic planning that incorporates environmental assessment in its development. However, such environmental considerations are less clear in broader policy areas such as energy or transport.

CONCLUSION: THE WAY AHEAD

This text attempts to provide the reader with an overview of Australian EIA procedures and includes some examples of how EIAs operate in practice. It has also attempted to place this in the broader global context of EIA development. In particular, this text has used the generally universal EIA processes of screening, scoping, documentation, public participation, assessment, decision-making, and monitoring to examine Australian EIA. Although this basic EIA structure is generally applicable around the world, it is complicated in Australia by the complexity of the legislative procedures and regulations, and their variation between the Commonwealth, the states, and the territories. In addition, the discussion here demonstrates that the practice of EIA sometimes takes on increased complexity through factors such as ad hoc procedures and guidelines, variations in political involvement, differences in formal and informal public participation, and confusion over the specificity of projects and levels of assessment.

For example, the Magnetic Quay proposals involved two jurisdictions and at least five pieces of legislation, was subject to three levels of EIA, and involved six separate reports plus an inquiry. Notwithstanding this high level of scrutiny, the project resulted in a partially completed development with significant environmental impacts, and the financial safeguards were inadequate for the task of rehabilitating the environment. In contrast, the Hindmarsh Island EIA procedures were relatively straightforward and were completed, in a short time-frame. However, in this case significant public and political involvement occurred after the formal EIA procedures had been completed and this involvement stalled the project before it proceeded to the second stage. While these two cases are extremes within the broader context of numerous routine and successful Australian EIAs, they do beg the question of why things go wrong and just how effective the EIA process is in Australia.

The answer to the question 'Why do things go wrong?' is often quite complicated, and for this reason it is essential to examine EIA in practice (as stressed in this text) to find some of the answers. Often the problems are quite specific to the particular location, such as Hindmarsh Island. This example may, however, also provide general lessons as to how future EIA procedures can better provide for formal public involvement of all stakeholders in a project before decisions are made. Other problems may occur because of political interference outside the formal EIA process. These are best dealt with by a broader environmental assessment of policies, plans, or programs (SEA) before the project-based EIA is undertaken. In some cases, problems with EIA projects can be generic. For example, in South Australia the decisions on a number of EIA projects were constrained because of incompatibility with the associated development plan. The short-term response was to use a different section of the legislation and a different decision maker. However, the longer-term response has been to address the problem in new EIA legislation.

There is general agreement that the Australian EIA process is effective in identifying and predicting potential environmental impacts and communicating this information to decision makers. Some differences of opinion exist in relation to the level of this effectiveness. For example, Porter (1987) comments that Australia has embraced the most worthwhile aspects of EIA and is ahead of other nations such as the United States and the United Kingdom. However, Woods's (1994) comparative study of eight EIA jurisdictions places the Australian system ahead of the United Kingdom but behind other systems such as those in the United States, the Netherlands, and Canada. It is interesting to note that out of all the eight EIA systems investigated, Woods gave the highest rating to Western Australia. This type of independent study suggests that the Australian EIA process is effective but has some weaknesses, such as the uncertainty caused by using discretionary criteria or the duplication and overlap of Commonwealth, state, and territory legislation.

In recognition of these criticisms, there have been some significant steps taken in the 1990s that provide a solid foundation for the future development of Australian EIA. These are:

- 1990 the recognition of the need for a cooperative approach to environmental matters generally, including EIA
- 1991 the endorsement of a national approach to EIA (*National Approach on EIA*)
- 1992 the provision of a foundation agreement on general principles of EIA through section 3 of the *Intergovernmental Agreement on the Environment*
- 1992 the drafting of
    - *Guidelines and Criteria for Determining the Need for and Level of EIA in Australia*
    - *National Agreement on EIA.*

  (Although both these documents took some time to endorse, the ad hoc acceptance of the *Guidelines and Criteria* has meant that they have been incorporated into some state legislation and regulations prior to their endorsement.)
- 1995 the endorsement of the *Guidelines and Criteria for Determining the Need for and Level of EIA in Australia*
- 1997 the endorsement of the *National Agreement on EIA*.

In addition, there have been a number of external influences on Australian EIA over this time. The Earth Summit conference in 1992 provided the impetus for Australia to present a coordinated national environmental approach—for example, the *Intergovernmental Agreement on the Environment*, and the ESD and Greenhouse Strategies. Notwithstanding Australia's subsequent loss of environmental credentials in the global forum, this earlier work laid a foundation for further improvements to EIA. Another important influence has been the number of key EIA practitioners and researchers who have become involved with EIA capacity building through:

- agencies such as UNEP and the United Nations Development Program
- the production of EIA training manuals
- the export of Australian expertise in this area.

There has also been an important exchange of ideas through tripartite (Australia, Canada, and New Zealand) workshops on EIA and through the involvement of the former Commonwealth Environmental Protection Agency in the international EIA effectiveness study.

Although some very positive steps have been taken towards improving the Australian EIA system, there remain a number of commonly identified weaknesses. These include:

- the lack of monitoring and auditing
- restricted appeal rights
- inadequate consideration of cumulative and trans-boundary effects
- the lack of quality control in relation to EIA consultants and EIA documentation.

Of these, monitoring and auditing perhaps warrant the most urgent attention.

The conclusions of the international EIA effectiveness study conducted by the Canadian Environmental Assessment Agency also have relevance for the future development of EIA in Australia. The main recommendations coming out of this study are recommendations to:

- establish standards for quality performance in EIA
- upgrade EIA processes and activities to improve quality control, public involvement, and cumulative effects
- transform SEA into an integral part of decision-making
- sharpen SEA as a tool for achieving sustainable development.

Provision for public involvement varies between the different EIA jurisdictions in Australia. The adequacy of such involvement is likely to be an ongoing issue as long as there are controversial projects.

Strategic environmental assessment appears to be one of the important issues confronting the future development of Australian EIA. The broad definition of EIA given at the start of this text includes environmental assessment of policies, plans, and programs, which indicates that SEA is part of EIA. Elsewhere in the world the environmental assessment of policies, plans, and programs are defined separately as SEA. This is significant in that SEA has an established formal approach in a number of other countries. It has been noted that, in effect, a form of SEA is conducted in Australia either in an ad hoc manner or subsumed under a different name such as strategic planning. There are currently moves afoot to consider a more formal approach to SEA in Australia, but how fast these changes proceed may well depend on the prioritisation of other issues, such as the development of the role of strategic planning.

The overriding imperative for EIA in Australia is to ensure proper environmental outcomes by predicting potential environmental impacts and communicating this information to decision makers. In addition to project-based EIA, there is clearly a need to conduct environmental assessment at the broader policy, plan, and program level. This may occur through broadening the scope of EIA, through the formalisation of SEA procedures, or through other mechanisms. There is currently an international push for environmental sustainability, and it appears that improved Australian environmental assessment procedures could be well placed as a tool for achieving such sustainability in this part of the globe.

Bibliography

Abeyratne, R. 1993, 'Sydney's Third Runway—a global perspective', *Environmental Planning and Law Journal*, October, vol. 10, pp. 355–63.

ACTPA. See Australian Capital Territory Planning Authority.

Adam, P. 1993, 'Botany Bay—The Third Runway and Some Issues in Environmental Assessment', *Australian Biologist*, vol. 6, no. 4, pp. 161–8.

Anderson, E. 1986, *Environmental Impact Assessment in Australia and Overseas*, report prepared for the Department of Arts, Heritage and Environment, Canberra.

—— 1994, Australian Country Paper, paper presented at the Seventh EIA Tripartite Workshop, 21–4 March, Canberra.

ANZECC. See Australian and New Zealand Environment Conservation Council.

Australian and New Zealand Environment and Conservation Council 1991a, *A National Approach to Environmental Impact Assessment in Australia*, ANZECC, Canberra.

—— 1991b, *Environmental Impact Assessment: Background Paper to the National Approach*, ANZECC, Canberra.

—— 1996, *Guidelines and Criteria for Determining the Need for and Level of Environmental Impact Assessment in Australia*, ANZECC, Canberra.

—— 1997, *National Agreement on Environmental Impact Assessment*, ANZECC, Canberra.

Australian Capital Territory Planning Authority 1992a, *Draft Environmental Impact Assessment Practices under the Land (Planning and Environment) Act 1991*, Department of Environment, Land and Planning, Canberra.

—— 1992b, *Proposed Plan for West Belconnen: Final Environmental Impact Statement*, ACTPA, Canberra.

Australian Groundwater Consultants and Koukouru & Partners 1988, *Wellington Cove Residential Community Resort, Draft Environmental Impact Statement*, prepared for Wellington Cove Pty Ltd.

—— 1989, *Wellington Cove Residential Community Resort: Supplement to the Draft Environmental Impact Statement*, prepared for Wellington Cover Pty Ltd.

Bailey, J. and English, V. 1991, 'Western Australian Environmental Impact Assessment: An Evolving Approach to Environmentally Sound Development', *Environmental and Planning Law Journal*, vol. 2, pp. 190–9.

—— 1994, 'Western Australian Environmental Impact Assessment: An Evolving Approach to Environmentally Sound Development', *Environmental and Planning Law Journal*, vol. 8, pp. 190–99.

Ballantyne, R. 1990, 'Runway: Councils Plan to Fight EIS', *Sydney Morning Herald*, 20 September, p. 1.

Battelle Institute 1978, *The Selection of Projects for EIA*, Commission of the European Communities Environment and Consumer Protection Service, Brussels.

Bates, G. 1992, *Environmental Law in Australia*, 3rd edn, Butterworths, Sydney.

—— 1994, 'Land Use Planning Reforms: Tasmania', *Environmental Planning and Law Journal*, vol. 11, no. 2, pp. 101–3.

—— 1995, *Environmental Law in Australia*, 4th edn, Butterworths, Sydney.

Beanlands, G. 1988, 'Scoping Methods and Baseline Studies in EIA', in P. Wathern (ed.), *Environmental Impact Assessment: Theory and Practice*, Routledge, London.

Beanlands, G. E. and Duinker, P. N. 1988, *An Ecological Framework for Environmental Impact Assessment in Canada*, Institute for Resource and Environmental Studies, Dalhousie University, Halifax, Nova Scotia.

Beckwith, J. A. 1994, 'Social Impact Assessment in Western Australia at a Cross Roads', *Impact Assessment*, vol. 12, no. 2, pp. 199–213

Binalong Pty Ltd 1989, *Draft Environmental Impact Statement for the Hindmarsh Bridge, Marina Extensions and Waterfront Development*, prepared by PPK Planning Consultants Pty Ltd, South Australia.

—— 1990, *Supplement to the Draft Environmental Impact Statement for the Hindmarsh Bridge, Marina Extensions and Waterfront Development*, prepared by PPK Planning Consultants Pty Ltd, South Australia.

Bisset, R. 1983, 'A Critical Survey of Methods for Environmental Impact Assessment in the United Kingdom', in T. O. O'Riordan and R. K. Turner (eds), *An Annotated Reader in Environmental Planning and Management*, Pergamon Press, Oxford.

—— 1988, 'Developments in EIA Methods', in P. Wathern (ed.), *Environmental Impact Assessment: Theory and Practice*, Routledge, London.

Bisset, R. and Tomlinson P. 1988, 'Monitoring and Auditing of Impacts', in P. Wathern (ed.), *Environmental Impact Assessment: Theory and Practice*, Routledge, London.

Bita, N. 1990, 'Council Plans to Test EIA Findings', *Sydney Morning Herald*, 20 September.

Bita, N. and Wilson, P. 1993, 'Tunnel May Cost an Extra $650 Million', *Australian*, 10 December, p. 4.

Blumer, D. 1993, 'Environmental Management on the Parallel Runway Construction Project—Sydney Airport', *Australian Biologist*, vol. 6, no. 4, pp. 175–9.

Bourman R. P. and Harvey, N. 1986, 'Landforms', in C. Nance and D. Speight (eds), *A Land Transformed: Environmental Change in South Australia*, Longman Cheshire, Melbourne, pp. 78–125.

Brew, D. 1995, 'General Introduction to EIA', in Alan Gilpin (ed.), *Environmental Impact Assessment (EIA): Cutting Edge for the Twenty-first Century*, Cambridge University Press, Cambridge.

Buckley, R. C. 1989, *Precision in Environmental Impact Prediction: First National Environmental Audit, Australia*, CRES/ANU Press, Canberra.

—— 1991, 'Environmental Planning Legislation: Court Backup Better than Ministerial Discretion', *Environmental Planning and Law Journal*, September, pp. 250–7.

Bureau of Industry Economics 1990, *Environmental Assessment—Impact on Major Projects*, Research Report no. 35, AGPS, Canberra.

Cameron McNamara Pty Ltd 1986, *Sydney Harbour Tunnel Environmental Impact Statement*, prepared for Transfield-Kumagai Joint Venture, New South Wales.

—— 1986, *Resort Concept Proposal, Wilpena Pound, Flinders Ranges*, South Australian Department of Tourism, Adelaide.

Canter, L. W. 1983, Environmental Impact Assessment: Current Status and Future Directions, paper delivered to the Symposium on Environmental Impact Assessment, Crete April 1983.

—— 1984, 'Environmental Impact Studies in the United States', in B. D. Clarke, A. Gilad, R. Bisset, and P. Tomlinson (eds), *Perspectives in Environmental Impact Assessment*, Reidel, Dordrecht, Netherlands.

Cardew, R. 1997, 'Development of Environmetal Study Programs', in *Directory of Postgraduate Study*, Australian Vice-Chancellor's Committee and Graduate Careers Council, Canberra.

CCNT. See Conservation Commission of the Northern Territory.

CEPA. See Commonwealth Environment Protection Agency.

Change Agency Pty Ltd 1995, *Review of the Commonwealth Environmental Impact Assessment Process*, prepared for the Commonwealth Environment Protection Agency.

Clark, B. 1983, *Aims and Objectives of Environmental Impact Assessment*, NATO [North Atlantic Treaty Organisation] Advanced Study Institute Series D, Behavioural and Social Sciences, no. 14, pp. 3–12.

—— 1995, 'Improving Public Participation in Environmental Impact Assessment', *Built Environment*, vol. 20, no. 4, pp. 294–308.

Clark, B., Bisset, R., and Wathern, P. 1980, *Environmental Impact Assessment*, Mansell, London.

CMT. See Copper Mines of Tasmania.

Cole, D. 1995, State and Territory Trends in Environmental Impact Assessment in Australia, paper delivered to the 14th National Environmental Law Association Conference, Sydney.

Commonwealth Environment Protection Agency 1991a, *Ecologically Sustainable Development Working Group on Agriculture*, *Final Report*, AGPS, Canberra.

—— 1991c, *Ecologically Sustainable Development Working Group on Energy Production*, *Final Report*, AGPS, Canberra.

—— 1991b, *Ecologically Sustainable Development Working Group on Energy Use*, *Final Report*, AGPS, Canberra.

—— 1991d, *Ecologically Sustainable Development Working Group on Fisheries*, *Final Report*, AGPS, Canberra.

—— 1991e, *Ecologically Sustainable Development Working Group on Forest Use*, *Final Report*, AGPS, Canberra.

—— 1991f, *Ecologically Sustainable Development Working Group on Manufacturing*, *Final Report*, AGPS, Canberra.

—— 1991g, *Ecologically Sustainable Development Working Group on Mining*, *Final Report*, AGPS, Canberra.

—— 1991h, *Ecologically Sustainable Development Working Group on Tourism*, *Final Report*, AGPS, Canberra.

—— 1991i, *Ecologically Sustainable Development Working Group on Transport*, *Final Report*, AGPS, Canberra.

—— 1992, *National Strategy for Ecologically Sustainable Development*, AGPS, Canberra.

—— 1994, *Public Review of the Commonwealth Environment Impact Assessment Process*, CEPA, Canberra.

—— 1995, A Guide to the Commonwealth Government's Procedures for Environmental Impact Assessment, <http://www.erin.gov.au/portfolio/epg/eianet.html> November 1997.

—— 1996, (unpublished) Proposed EIA Reform Package, as of 23 January 1996.

Commonwealth of Australia 1990, *Ecologically Sustainable Development: A Commonwealth Government Discussion Paper*, AGPS, Canberra.

Conservation Commission of the Northern Territory 1984, *Tindal RAAF Base Development: Environmental Assessment Report*, CCNT, Darwin.

—— 1990, *A Guide to the Environmental Assessment Process in the Northern Territory*, CCNT, Darwin.

Cook, D. and Voumard, S. 1994, 'The Tunnel Turns Two—and We're Still $3m in the Red', *Sydney Morning Herald*, 31 August, p. 3.

Coopers & Lybrand 1994, *An Analysis of EIA Practices and Procedures in Australian States and Territories*, paper prepared for the Commonwealth Environment Protection Agency Review of Environmental Impact Assessment, Commonwealth of Australia, Canberra.

—— 1995, *Copper Mines of Tasmania: Mt Lyell Mine Redevelopment Environmental Management Plan*, Thompson & Brett Consulting Engineers, Tasmania.

Co-ordinator General, Premier's Department 1987, *Impact Assessment in Queensland: Policies and Administrative Arrangements*, Co-ordinator General, Premier's Department, Brisbane.

Copper Mines of Tasmania 1995, *Mt Lyell Mine Redevelopment: Environmental Management Plan*, prepared by Thompson & Brett Consulting Engineers Pty Ltd, Tasmania.

Coultan, M. 1988, 'Greiner Weighs Up Options on Tunnel', *Sydney Morning Herald*, 3 May, p. 2.

—— 1994, 'Public May Foot Tunnel Tax Bill', *Sydney Morning Herald*, 19 October, p. 48.

Council on Environmental Quality 1979, *Tenth Annual Report*, United States Government Printing Office, Washington DC.

Damania, D. 1992, 'Environmental Policy', in A. Parkin and A. Patience (eds), *The Bannon Decade: The Politics of Restraint in South Australia*, Allen & Unwin, Sydney.

DASETT. See Department of Arts, Sport, the Environment, Tourism and Territories (Commonwealth).

Davies, G. S. and Muller, F. G. 1983, *A Handbook for Environmental Impact Assessment for Use in Developing Countries*, The United Nations Environment Programme, Nairobi.

Dee, N., Baker, J. K., Drobny, N. L., Duke, K. M., Whitman, I., and Fahringer, D. C. 1973, 'An Environmental Evaluation System for Water Resources Planning', *Water Resources Research*, vol. 3, pp. 523–55.

De Jongh, P. 1988, 'Uncertainty in EIA', in P. Wathern (ed.), *Environmental Impact Assessment: Theory and Practice*, Routledge, London.

DELM (Tas). See Department of Environment and Land Management (Tasmania).

DEP (NSW). See Department of Environment and Planning (New South Wales).

DEP (WA). See Department of Environmental Protection (Western Australia).

Department of Arts, Sport, the Environment, Tourism and Territories (Commonwealth) 1988, *Assessment Report for the Magnetic Quay Development, Nelly Bay, Magnetic Island*, DASETT, Canberra.

—— 1991, *Proposed Third Runway, Sydney (Kingsford-Smith) Airport, Environmental Assessment Report*, DASETT, Canberra.

Department of Environment and Land Management (Tasmania) 1994, *Assessment of New Development Proposals in Tasmania: A Summary of Procedures in Use as at May 1994*, DELM (Tas), Hobart.

—— 1996, *Environmental Assessment Manual: Version 1*, Division of Environmental Management, DELM (Tas), Hobart.

Department of Environment and Planning (New South Wales) 1987, *Proposed Sydney Harbour Tunnel: Environmental Impact Assessment*, DEP (NSW), Sydney.

Department of Environmental Protection (Western Australia) 1996, *Auditing Environmental Conditions: Guidelines for Proponents*, Environmental Audit Branch, DEP (WA), Perth.

—— 1993/4, *Annual Report*, DEP (WA), Perth.

Department of Home Affairs and the Environment (Northern Territory) 1984, *Environmental Assessment Report: RAAF Base Tindal, Northern Territory*, DHAE (NT), Darwin.

Department of Housing and Urban Development (South Australia) 1997, *Guide to the Assessment of Major Developments of Projects: An Integrated Planning and Development Assessment System for South Australia*, DHUD (SA), Planning Division: Adelaide.

Department of Housing, Local Government and Planning (Queensland) 1995, Planning, Environment and Development Assessment Bill: Exposure Draft for Public Comment–Additional Material, DHLGP (Qld), Brisbane.

Department of Lands, Planning and Environment (Northern Territory) 1995, *Environmental Assessment in the Northern Territory*, DLPE, Environment Protection Division, Darwin.

Department of Main Roads (New South Wales) 1987, *Sydney Harbour Tunnel: Report on Environmental Impact Assessment*, DMR (NSW), Sydney.

Department of Planning (New South Wales) 1991, *Modernising the Planning System in NSW: Discussion Paper*, Department of Planning, Sydney.

—— 1985, *EIA Manual*, Department of Planning, Sydney.

—— 1994, *Annual Report*, Department of Planning, Sydney.

Department of Planning and Development (Victoria) 1993, *Assessment and Panel Report, Mornington Boating Facilities Study*, Minister for Planning, DPD (Vic), Melbourne.

—— 1995, *Guidelines for Environmental Impact Assessment and the Environment Effects Act*, DPD (Vic), Melbourne.

Department of the Premier, Economic and Trade Development (Queensland) 1995, *State Development and Public Works Organisation Amendment Act and Regulation: Exposure Draft for Public Comment*, DPETD (Qld), Brisbane.

DHAE (NT). See Department of Home Affairs and Environment (Northern Territory).

DHLGP(Qld). See Department of Housing, Local Government and Planning (Queensland).

DHUD (SA). See Department of Housing and Urban Development (South Australia).

DLPE. See Department of Lands, Planning and Environment (Northern Territory).

DMR (NSW). See Department of Main Roads (New South Wales).

DPD (Vic). See Department of Planning and Development (Victoria).

DPETD (Qld). See Department of the Premier, Economic and Trade Development (Queensland).

Edmonds V. A. 1989, *An Archeological Survey of the Marina Goolwa, Hindmarsh Island*, Binalong Pty Ltd, South Australia.

EIA Review Committee 1987, *Environmental Impact Assessment in South Australia*, Department of Environment and Planning, South Australia, Adelaide.

ENERSOL Consulting Engineers 1989, *Sydney Harbour Tunnel Inquiry: Vols. 1 and 2*, prepared for North Sydney Municipal Council, ENERSOL, Sydney.

England, P. 1996, 'Change of Form or Change of Philosophy? A Critical Review of the Queensland PEDA Bill', *Environmental and Planning Law Journal*, vol. 13, no. 2, pp. 133–45.

Environment Assessment Branch, Department of Planning and Development (Victoria) 1991, *Environmental Assessment in Victoria*, Ministry for Planning and Environment, Melbourne.

—— 1995, 'Environmental Assessment in Victoria', in *Future Directions for EIA in Victoria: Consultative Committees and Other Initiatives: Background Papers*, DPD, and Department of Planning, Policy and Landscape, Melbourne.

Environment Institute of Australia 1990, Environmental Impact Assessment Processes in Australia: A Discussion Paper, presented at the National Workshop on Environmental Impact Assessment (EIA), October 1990, (place of conference unavailable).

Environmental Protection Authority (Western Australia) 1991, *Proposed Quicklime Plant and Limestone Quarry at Nowergup, Near Wanneroo*, (report and recommendations of the Environmental Protection Authority), Bulletin 584, Assessment No. 491, DEP (WA), Perth.

—— 1993, *Environmental Reviews: Guidelines For Proponents*, DEP (WA), Perth.

—— 1993–4, *Annual Report*, DEP (WA), Perth.

—— 1994, 'Quicklime Plant and Limestone Quarry, Nowergup near Wanneroo—Change of Environmental Conditions to Allow Increases in Production Rates: Report and Recommendations of the Environmental Protection Authority', *Bulletin 733*, DEP (WA), Perth.

—— 1994–5, *Annual Report*, DEP (WA), Perth.

——1995, *Environmental Impact Assessment: How To Make a Submission*, DEP (WA), Perth.

—— 1996, *Auditing Environmental Conditions: Guidelines for Proponents*, DEP (WA), Perth.

EPA(WA). See Environmental Protection Authority (Western Australia).

Evans, E. 1976, An Examination of the EIS Technique and its Role in the Planning Process, unpublished thesis, Master of Urban and Regional Planning, University of Adelaide.

FAC. See Federal Airports Commission.

Fairfax, S. K. and Ingram, H. M. 1981, 'The United States Experience', in T. O'Riordan and W. R. D. Sewell (eds), *Project Appraisal and Policy Review*, Wiley & Sons, Chichester, United Kingdom.

Farrier, D. 1993, 'Commonwealth Environmental Impact Assessment Legislation', in *Environmental Law Handbook: Planning and Land Use in New South Wales*, 2nd edn, Redfern Legal Centre Publishing, Sydney.

Federal Airports Commission 1990, *Proposed Third Runway, Sydney (Kingsford-Smith) Airport: Draft Environmental Impact Statement*, prepared by Kinhill Engineers Pty Ltd for Federal Airports Commission, Sydney.

—— 1991, *Proposed Third Runway, Sydney (Kingsford-Smith) Airport: Supplement to the Draft Environmental Impact Statement*, prepared by Kinhill Engineers Pty Ltd for Federal Airports Commission, Sydney.

Fookes, T. 1987a, 'The Integration of EIA and Planning: The South Australian Experience', in A. Conacher (ed.), *Readings in Australian Geography: Proceedings of the 21st Institute of Australian Geographers' Conference*, University of Western Australia, Perth, May 1986.

—— 1987b, 'A Comparison of Environmental Impact Assessment in South Australia and Proposed United Nations Environment Programme Goals and Principles', *Environmental and Planning Law Journal*, vol. 4, pp. 204–15.

Formby, J. 1981, 'The Australia Experience', in T. O'Riordan and W. R. D. Sewell (eds), *Project Appraisal and Policy Review*, John Wiley & Sons, Chichester, United Kingdom.

—— 1987, 'The Australian Government's Experience with Environmental Impact Assessment', *Environmental Impact Assessment Review*, no. 7, pp. 207–26.

Fowler, R. 1982, *Environmental Impact Assessment, Planning and Pollution Measures in Australia*, prepared for the Department of Home Affairs and Environment, AGPS, Canberra.

Gardner, A. 1995, 'The Planning Legislation Amendment Bill 1994 (WA) — implications for EIA in WA', *Environmental and Planning Law Journal*, February, vol. 12, no. 1, pp. 10–15.

GBRMPA. See Great Barrier Reef Marine Park Authority.

Gibson, R. B. 1993, 'Environmental Assessment Design: Lessons from the Canadian Experience', *The Environmental Professional*, vol. 15, pp. 12–24.

Gilpin, A. 1995, *Environmental Impact Assessment (EIA): Cutting Edge for the Twenty-First Century*, Cambridge University Press, Cambridge.

Giroult, E. 1988, 'WHO Interest in Environmental Health Impact Assessment', in P. Wathern (ed.), *Environmental Impact Assessment, Theory and Practice*, Unwin Hyman, London.

Glasson, J., Elson M. J., Van der Wee, M., and Barrett, B. 1987, *Socio-economic Impact Assessment: Theory and Practice*, Routledge, London, pp. 62–84.

Glasson, J., Therivel, R., and Chadwick, A. 1994, *Introduction to Environmental Impact Assessment*, University College London Press, London.

Government of Western Australia 1992, *Review of the Environmental Protection Act 1986*, Report of the Independent Advisory Committee for the Review of the Environmental Protection Act to the Minister for the Environment, Western Australia.

Great Barrier Reef Marine Park Authority 1988, *Assessment Report: Magnetic Quay Marina Proposal*, Magnetic Keys Limited, Townsville.

—— Training Package Using the Magnetic Quays/Nelly Bay Proposal as a Working Example, courtesy Clive Cook, Management Impact Assessment and Major Projects, Environmental Impact Management, GBRMPA.

Grosse, C. 1990, 'Beach "Doomed" Under Marina Plan', *Mornington Leader*, 3 October.

Hare, W. (ed.) 1990, *Ecologically Sustainable Development*, Australian Conservation Foundation, Greenpeace (Australia) with the Wilderness Society and the World Wildlife Fund for Nature.

Hare, W. 1992, 'Achieving International Environmental Objectives', *Habitat Australia*, February, pp. 12–14.

Harvey, N. 1992a, 'South Australia's Coastal Marina Strategy: Planning Success or Recession Victim?', *Australian Planner*, vol. 30, no. 1, pp. 4–7.

—— 1992b, 'The Relationship Between Environmental Impact Assessment and Ecologically Sustainable Development in Australia: A Critique of Recent National Reports', *Environmental and Planning Law Journal*, vol. 9, no. 4, pp. 265–73.

—— 1993a, 'Sensitivity of South Australian Environments to Marina Construction', in R. J. Allison and D. S. G. Thomas (eds), *Environmental Sensitivity*, John Wiley & Sons, Chichester, United Kingdom.

—— 1993b, 'Variability of Coastal Engineering Data in South Australian Environmental Impact Assessment 1982–93', *Proceedings of the Eleventh Australasian Conference on Coastal and Ocean Engineering*, Institution of Engineers, Australia, pp. 487–94.

—— 1994a, 'Timing of Environmental Impact Assessment: Where Are the Delays?', *Australian Planner*, vol. 13, no. 3, pp. 125–30.

—— 1994b, 'Energy Related Projects and Environmental Impact Legislation in South Australia', *Energy & Environment*, vol. 5, no. 4, pp. 285–303.

—— 1995, *Environmental Impact Statements for Major Projects in South Australia: 1982–1994*, South Australian Geographical Papers, no. 4, Royal Geographical Society of Australasia Inc. (South Australian Branch), South Australia.

—— 1996, 'Public Involvement in the Environmental Impact Assessment Process: The Hindmarsh Island Bridge Case', *Australian Planner*, vol. 33, pp. 36–43.

Harvey, N. and Ferguson, K. 1994, 'Environmental Impact Assessment in South Australia: Towards 2000', University of South Australia and South Australian Planning Education Foundation, Planning Paper Series, no. 4, University of South Australia, Adelaide.

Harvey, N. and Swift, P. 1990a, 'Contrasts in Processing Coastal Marina Applications in South Australia', *Australian Planner*, vol. 28, no. 3, pp. 44–7.

—— 1990b, 'Coastal Marinas in South Australia: Environmental Issues and Strategic Planning', *Australian Geographer*, vol. 21, no. 2, pp. 141–50.

Harvey, T. 1981, 'Environmental Interventions: The Monitoring Paradigm I., The Monitoring Concept and Practices of Descriptive Monitoring', *The Environmentalist*, vol. 1, pp. 283–91.

Haydon, J. 1991, 'Major Shopping Developments and their Environmental Impact Assessment in Queensland', *Environmental Planning and Law Journal*, April, pp. 115–29.

Healy, R. and Childs, L. 1994, Response to CEPAs Initial Discussion Paper (Setting the Direction), public review of the Commonwealth environmental impact assessment process, prepared for the Wildlife Preservation Society of Queensland, and the Byfield Residents Action Group.

—— 1994a, Response to Administrative Review Council Discussion Paper (Review of Commonwealth Environmental Impact Assessment Decisions), prepared for the Wildlife Preservation Society of Queensland, and the Byfield Residents Action Group.

Heer, J. E. and Hagerty, D. J. 1977, *Environmental Assessments and Statements*, Van Nostrand Reinhold, New York.

Hill, T. 1992, 'Relationship Between Commonwealth and State Environmental Assessment Statutes', *Australian Environmental Law News: The Journal of the National Environmental Law Association Limited*, no. 4, pp. 21–3.

Hollick, M. 1981, 'EIA and Environmental Management in Western Australia', *Environmental Impact Assessment Review*, vol. 2, no. 1, pp. 116–20.

ICESD. See Intergovernmental Committee for Ecologically Sustainable Development (Commonwealth).

IGAE. See *Intergovernmental Agreement on the Environment.*

Intergovernmental Agreement on the Environment 1992, AGPS, Canberra.

Intergovernmental Committee for Ecologically Sustainable Development (Commonwealth) 1996, *Report on the Implementation of the National Strategy for Ecologically Sustainable Development, 1993–1995*, Canberra.

International Union for the Conservation of Nature, United Nations Environment Programme and World Wildlife Fund for Nature 1980, *World Conservation Strategy*, Geneva.

IUCN. See International Union for the Conservation of Nature.

Jenkins, B. 1988, 'RAAF Base Tindal Environmental Studies', *International Association for Impact Assessment VII Annual Meeting: Abstracts*, 5–9 July, International Association for Impact Assessment, Griffith University, Brisbane.

—— 1990. 'Australian Planning and Assessment Techniques in Relation to Sustainable Development', in Kinhill Engineers Pty Ltd and Phillips Fox Solicitors and Attorneys, *Approval Systems for Major Projects in Australia: A Comparative Review*, report prepared for the Commonwealth Department of Industry, Technology and Commerce, Canberra.

—— 1993, 'Sydney Airport Third Runway EIS–Coverage of Marine Ecology Issues', *Australian Biologist*, vol. 6, no. 4, pp. 190–4.

Jones, C. E., Lee, N., and Wood, C. 1991, *UK Environmental Impact Statements, 1988–90: An Analysis*, Department of Planing and Landscape, University of Manchester, United Kingdom.

Jubilee Point Pty Ltd 1986a, *Jubilee Point Development Project, Draft Environmental Impact Statement*, prepared by Kinhill Stearns Pty Ltd, Kinhill Stearns, South Australia.

—— 1986b, *Jubilee Point Development Project, Supplement to the Draft Environmental Impact Statement*, prepared by Kinhill Stearns Pty Ltd, Kinhill Stearns, South Australia.

Kennedy, W. V. 1986, *Environmental Impact Assessment and Highway Planning: A Comparative Case Study Analysis of the United States and the Federal Republic of Germany*, Raner Bohn Verlag, Science Centre, Berlin, Germany.

—— 1988, EIA in Global Perspective: Does it Work, paper presented to Environmental Impact Assessment as a Management Tool Workshop, November 1987, Canberra.

Kinhill Engineers Pty Ltd 1991, *Swan Portland Cement Limited, Proposed Quicklime Project at Nowergup: Consultative Environmental Review*, Kinhill Engineers Pty Ltd, Western Australia.

—— 1993, *Swan Portland Cement Limited Nowergup Quicklime Project: Referral of a Revised Proposal*, Kinhill Engineers Pty Ltd, Western Australia.

—— 1994, *Public Participation in the Environmental Impact Assessment Process*, input to the public review of the Commonwealth Environmental Impact Assessment process, Canberra.

Kinhill Stearns 1983, *RAAF Base Tindal: Draft Environmental Impact Statement*, prepared by Kinhill Stearns for Department of Defence, Northern Territory.

Kinnaird, M. 1987, The Development Approval Process—Lessons for the Future, unpublished copy of a speech to the Real Estate Professionals Convention, November 1987, Adelaide.

—— 1989, Keynote Address, unpublished copy of a speech to the 1989 Annual Conference of the Australasian Institute of Mining and Metallurgy, May 1989, Perth.

Kriwoken, L. 1997, 'Current EIA Practice in Tasmania', in N. Harvey and M. McCarthy (eds), *Environmental Impact Assessment for the 21st Century*, Conference Proceedings, 9–11 April, Mawson Graduate Centre for Environmental Studies, University of Adelaide, South Australia.

Krohn, J. 1997, 'Current EIA Practice in Victoria', in N. Harvey, and M. McCarthy (eds), *Environmental Impact Assessment for the 21st Century, Conference Proceedings*, 9–11 April, Mawson Graduate Centre for Environmental Studies, University of Adelaide, South Australia.

Leadbitter, D. and McDonall, V. 1993, 'EIA and Botany Bay: Lessons for the Fishing Industry', *Australian Biologist*, vol. 6, no. 4, pp. 169–74.

Lee, N., and Colley, R. 1990, *Reviewing the Quality of Environmental Statements*, Department of Planning and Landscape, University of Manchester, United Kingdom.

'Legislative Update: Commonwealth—New Directions in the Commonwealth Environmental Impact Assessment Process' 1995, *Environment Institute of Australia Newsletter*, no. 29, pp. 32–4.

'Legislative Update: Northern Territory Environmental Assessment Act 1994' 1995, *Environment Institute of Australia Newsletter*, no. 29, pp. 44–5.

Leopold, L. B., Clarke, F. E., Hanshaw B. B., and Balsey J. R. 1971, 'A Procedure for Evaluating Environmental Impact, Washington DC', *US Geological Survey Circular* 645.

McCarthy, M. 1996, 'Strategic Environmental Assessment: The Potential for Development in South Australia', *Project Appraisal*, vol. 11, no. 3, pp. 146–56.

McDonald, J. 1995, 'One Step Forward, Two Steps Back: Environmental Impact Assessment Reform Proposals in Queensland', *Environmental and Planning Law Journal*, vol. 12, no. 6, pp. 367–72.

McHarg, I. 1968, *A Comprehensive Route Selection Method*, Highway Research Record No. 246, Highway Research Board, Washington DC.

McIntyre & Associates Pty Ltd 1988a, *Magnetic Quay, Nelly Bay — Magnetic Island: Public Environment Report*, prepared for Magnetic Keys Limited, Queensland.

—— August 1988b, *Magnetic Quay, Nelly Bay — Magnetic Island: Public Environment Report*, prepared for Magnetic Keys Limited, Queensland.

McNamara 1986, *Sydney Harbour Tunnel Environmental Impact Statement*, prepared for Transfield-Kumagai Joint Venture, New South Wales.

Mant, J. H. 1982, 'Development Control: A Tale of Three Acts', *Australian Planner*, vol. 20, no. 2, pp. 74–6.

MBFWG. See Mornington Boating Facilities Working Group.

Milner, G. and Mazumder, M. R. 1993, 'Environmental Management of the Impacts of the Parallel Runway on Botany Bay', *Australian Biologist*, vol. 6, no. 4, pp. 180–9.

MKL. See McIntyre & Associates.

Mornington Boating Facilities Working Group 1990, *Mornington Boating Facilities Study and Environment Effects Statement*, prepared by Loder & Bayly Consulting Group for the Mornington Boating Facilities Working Group, Victoria.

Morris, L. 1994, 'Tunnel Payout Climbs to $4bn', *Sydney Morning Herald*, 17 October, p. 3.

Morris, L. and Therivel, R. 1995, *Methods of Environmental Impact Assessment*, University College London Press, London.

Morrison, M. D., Blamey, R. K., Bennett, J. W., and Louviere, J. J. 1996, *A Comparison of Stated Preference Techniques for Estimating Environmental Values*, Modelling Research Report No. 1, The University of New South Wales, Sydney.

Münchenberg, S. 1994a, 'Judicial Review and the Commonwealth Environment Protection (Impact of Proposals) Act 1974', *Environmental and Planning Law Journal*, vol. 11, no. 6, pp. 461–78.

—— 1994b, 'Review of the Commonwealth Environmental Impact Assessment Process', *Environmental and Planning Law Journal*, vol. 11, no. 6, p. 460.

—— 1995, 'Amendments to Commonwealth EIA Procedures', *Environmental and Planning Law Journal*, vol. 12, no. 4, pp. 235–7.

Munn, R. E. 1979, *Environmental Impact Assessment: Principles and Procedures*, SCOPE report 5, 2nd edn, John Wiley & Sons, Chichester, United Kingdom.

NBH. See North Barrier Holdings.

New South Wales Science and Technology Council 1983, 'The Science and Technology of Environmental Impact Assessment, with Particular Reference to Methods Used in New South Wales', *Federal Environmental Assessment Review Office* [Canada] *Report*, Federal Environnmetal Assessment Review Office, Canada.

Norman, B. 1997, 'Current EIA Practice in the Australian Capital Territory', in N. Harvey and M. McCarthy (eds), *Environmental Impact Assessment for the 21st Century, Conference Proceedings*, 9–11 April, Mawson Graduate Centre for Environmental Studies, University of Adelaide, South Australia, pp. 83–4.

North Barrier Holdings 1986, *Magnetic Keys Development, Nelly Bay, Magnetic Island*, 2 vols, prepared by McIntyre & Associates for North Barrier Holdings Ltd, Queensland.

Office of Planning and Urban Development (South Australia) 1992, *MFP Australia Gillman—Dry Creek Urban Development Proposal Assessment Report*, Assessments Branch, OPUD, Adelaide.

OPUD. See Office of Planning and Urban Development (South Australia).

O'Riordan, T. 1979, 'Environmental Impact Analysis and Risk Assessment in a Management Perspective', in G. T. Goodman and W. D. Rowe (eds), *Energy Risk Management*, Academic Press, London.

O'Riordan, T. and Sewell, W. R. D. (eds) 1981, *Project Appraisal and Policy Review*, John Wiley & Sons, Chichester, United Kingdom.

Panel Report 1991, Panel Report on Amendment L28 to the Shire of Mornington, Planning Scheme and the Environment Effects Statement on the Proposed Boating Facilities Development, Victoria.

Parkin, A. and Patience. A (eds) 1992, *The Bannon Decade: The Politics of Restraint in South Australia*, Allen & Unwin, Sydney.

Porter, C. F. 1985, *Environmental Impact Assessment: A Practical Guide*, University of Queensland Press, St Lucia, Queensland.

—— 1987, 'Environmental Impact Assessment', in T. Hundloe and R. Neumann (eds), *Environmental Practice in Australia*, Environmental Institute of Australia, Canberra.

RAC. See Resource Assessment Commission (Commonwealth).

Raff, M. 1995, 'The Renewed Prominence of Environmental Impact Assessment: "A Tale of Two Cities"', *Environmental and Planning Law Journal*, vol. 12, no. 4, pp. 241–63.

Rau, J. G. and Wooten, D. C. 1980, *Environmental Impact Analysis Handbook*, McGraw-Hill, New York.

Redfern Legal Centre 1993, *The Environmental Law Handbook: Planning and Land Use in New South Wales*, 2nd edn, Redfern Legal Centre Publishing, Sydney.

Rees, W. E. 1988, 'A Role for Environmental Assessment in Achieving Sustainable Development', *Environmental Impact Assessment Review*, vol. 8, pp. 273–91.

Remove All Rubbish Company 1983, *Draft Environmental Impact Statement of the Proposed Solid Waste Depot*, prepared by Wallman Planning Consultants Pty Ltd, Adelaide.

Resource Assessment Commission 1993, *Coastal Zone Inquiry*, RAC, Commonwealth of Australia, Canberra.

Rissel, C. 1990, 'The Health Effects of a Third Runway at Kingsford–Smith Airport', *New Doctor*, no. 52, summer, pp. 15–18.

—— 1991, 'The Limitations of Sydney Airport's Third Runway Environmental Impact Statement', *New Doctor*, no. 55, autumn, pp. 14–16.

—— 1992, 'Principles and Ethics in Making Decisions About Environmental Impact Statements', *New Doctor*, no. 57, autumn, pp. 23–5.

Robinson, D. 1994, 'Public Participation in Decision-Making', *Environmental Planning Law Journal*, pp. 320–37

SADEP. See South Australian Department of Environment and Planning.

Sadler, B. 1996, *Environmental Assessment in a Changing World: Evaluating Practice to Improve Performance: Final Report of the International Study of the Effectiveness of Environmental Assessment*, prepared for the Canadian Environmental Assessment Agency and International Association for Impact Assessment, Canada.

Saddler, K. 1990, 'Group Vows to Fight Them on the Beaches', *Mornington Leader*, 19 September.

Sammarco, P. W. 1990, The Resource Assessment Commission, and Environmental Impact Assessment in Australia: Major Issues, unpublished paper presented to Environmental Impact Assessment in Australia National Conference, Environmental Institute of Australia, 25 October.

SANPS. See South Australia National Parks and Wildlife Service.

Saunders, R. 1995, 'The Engineer's Role in Community Consultation', in *Future Directions for EIA in Victoria: Consultative Committees and Other Initiatives*, Background Papers, Department of Planning and Development and Department of Planning, Policy and Landscape, Melbourne.

Scanlen, T. 1993, Current Requirements for Impact Assessment of Property Developments in Queensland, paper presented at the Queensland Environmental Law Association Annual Conference, May 1993, Townsville.

SDAC. See Sustainable Development Advisory Council.

Sinclair Knight Merz 1995, *Nelly Bay Harbour: Environmental Impact Statement, Draft*, vols 1 and 2, prepared for the Department of the Premier, Economic and Trade Development, Brisbane.

Smyth, R. B. 1987, The Environmental Planning and Assessment Act as a Management Tool, paper presented to a seminar on environmental legislation and its impact on management, The Institution of Engineers Australia, Department of Environment and Planning, Sydney.

Snashall, D. 1993, 'Local Government and Environmental Impact Assessment', *Australian Environmental Law News*, no. 4, pp. 29–37.

South Australia National Parks and Wildlife Service 1988a, *Proposed Wilpena Station Resort, Flinders Ranges National Park, Draft Amendment to the Flinders Ranges National Park Management Plan and Draft Environmental Impact Statement*, prepared by Michael Williams and Associates for South Australia Parks and Wildlife Service.

—— 1988b, *Proposed Wilpena Station Resort, Flinders Ranges National Park, Supplement to the Draft Environmental Impact Statement*, prepared by Michael Williams and Associates for South Australia Parks and Wildlife Service.

South Australian Department of Environment and Planning 1990, *Assessment Report for the Proposed Hindmarsh Island Bridge Marina Extensions and Waterfront Development*, Major Projects and Assessments Branch, SADEP, Adelaide.

—— 1982, *Procedures Manual for Development Control*, SADEP, Adelaide.

Stone, Y. 1997, 'Current EIA Practice in New South Wales', in N. Harvey and M. McCarthy (eds), *Environmental Impact Assessment for the 21st Century, Conference Proceedings*, 9–11 April, Mawson Graduate Centre for Environmental Studies, University of Adelaide, South Australia, pp. 83–4.

Sustainable Development Advisory Council August 1995a, *CMT Project, Queenstown, Tasmania: Final Report*, SDAC, Hobart.

—— 1995b, *Draft Assessment Report: Copper Mines of Tasmania Project*, SDAC, Hobart.

Sykes, T. 1994, 'Politician's Heads Buried in Sand', *Bulletin*, 21 June, pp. 22–3.

Tatman, C. 1990, 'Harbor "Death Trap" Charge', *Mornington Leader*, 26 September.

Therivel, R. and Partidaro, M. R. 1996, *The Practice of Strategic Environmental Assessment*, Earthscan, London.

Therivel, R., Wilson, E., Thompson, S., Heaney, D., and Pritchard, D. 1992, *Strategic Environmental Assessment*, Earthscan, London.

Thomas, I. 1987, *Environmental Impact Assessment: Australian Perspectives and Practice*, Graduate School of Environmental Science, Monash University, Melbourne.

—— 1995, 'Background to EIA', in *Future Directions for EIA in Victoria: Consultative Committees and Other Initiatives*, Background Papers, Department of Planning and Development and Department of Planning, Policy and Landscape, Melbourne.

—— 1996, *Environmental Impact Assessment in Australia: Theory and Practice*, Federation Press, Sydney.

Thompson & Brett Consulting Engineers 1995, *Copper Mines of Tasmania Mt Lyell Mine Redevelopment: Environmental Management Plan*, prepared for Copper Mines of Tasmania, Tasmania.

Tomlinson, P. 1984, 'The Use of Methods in Screening and Scoping', in B., D. Gilad, R. Bisset, and P. Tomlinson (eds), *Perspectives in Environmental Impact Assessment*, Reidel, Dordrecht, Netherlands.

—— 1989, 'Environmental Statements: Guidelines for Review and Audit', *The Planner*, vol. 75, no. 28, pp. 12–15.

Tomlinson, P. and Atkinson, S. F. 1987, 'Environmental Audits: A Literature Review', *Environmental Monitoring and Assessment*, vol. 8, pp. 239–61.

Tongcumpou, C. and Harvey, N. 1994, 'Implications of Recent EIA Changes in Thailand', *Environmental Impact Assessment Review*, vol. 14, no. 4, pp. 271–94.

Totaro, P. 1994, 'Auditor Criticises Secrecy on Public Works Contracts', *Sydney Morning Herald*, 18 October, p. 1.

UN. See United Nations.

Underwood, T. 1993, 'How Not to Design an Environmental Monitoring Program: A Case Study From the FAC (Up in Botany Bay)', *Australian Biologist*, vol. 6, no. 4, pp. 194–97.

United Nations 1992, 'Rio Declaration on Environment and Development, 1992', in *Agenda 21*, United Nations, Rio de Janeiro, Brazil.

Wathern, P. (ed.) 1988, *Environmental Impact Assessment, Theory and Practice*, Unwin Hyman, London.

WCED. See World Commission on Environment and Development.

Weir, M. J. 1994, 'Putting the Environment into the Queensland Local Government (Planning and Environment) Act', *Environmental Planning and Law Journal*, vol. 11, no. 5, October, pp. 409–18.

Western Mining Corporation 1983a, *Kingston Lignite Project, Draft Environmental Impact Statement*, prepared by Kinhill Stearns for Western Mining Corporation, Adelaide.

—— 1983b, *Kingston Lignite Project: Supplement to the Draft Environmental Impact Statement*, prepared by Kinhill Stearns for Western Mining Corporation, Adelaide.

Whitehouse J. F. 1992, *Review of the Magnetic Island Marina Development*, Department of the Arts, Sport, the Environment and Territories, AGPS, Canberra.

Whyalla District Council 1986, *Whyalla Boat Harbour, Draft Environmental Impact Statement*, prepared by Kinhill Stearns Pty Ltd for Whyalla District Council, Adelaide.

—— 1987, *Whyalla Boat Harbour: Supplement to the Draft Environmental Impact Statement*, prepared by Kinhill Stearns Pty Ltd forWhyalla District Council, Adelaide.

Wilkins, L. and Rissel, C. E. 1990, 'Health Effects of a Third Runway Part 2: An Environmental Impact Statement—As If People Mattered', *New Doctor*, no. 53, winter, pp. 26–9.

Wood, C. 1992, 'Strategic Environmental Assessment in Australia and New Zealand', *Project Appraisal*, no. 7, pp. 143–9.

—— 1993, 'Environmental Impact Assessment in Victoria: Australian Discretion Rules EA!', *Journal of Environmental Management*, no. 39, pp. 281–95.

—— 1994, 'Evaluating Impact Assessment Systems', *Integrated Environmental Management*, vol. 30, pp. 10–13.

—— 1995, *Environmental Impact Assessment: A Comparative Review*, Longman Scientific & Technical, United Kingdom.

Wood, C. and Bailey, J. 1994, 'Predominance and Independence in Environmental Impact Assessment: the Western Australia Model', *Environmental Impact Assessment Review*, vol. 14, pp. 37–59.

World Commission on Environment and Development 1987, *Our Common Future*, (commonly known as 'the Bruntland Report'), Oxford University Press, Oxford.

Index